MONTANA'S
WALDRON CREEK FIRE

MONTANA'S WALDRON CREEK FIRE

THE 1931 TRAGEDY AND THE FORGOTTEN FIVE

DR. CHARLES PALMER

THE History PRESS

Published by The History Press
Charleston, SC
www.historypress.net

First published 2015

Manufactured in the United States

ISBN 978.1.46711.926.9

Library of Congress Control Number: 2015948833

Notice: The information in this book is true and complete to the best of our knowledge. It is offered without guarantee on the part of the author or The History Press. The author and The History Press disclaim all liability in connection with the use of this book.

CONTENTS

Preface

THE BACKDROP

In 2003, when I first encountered Nancy Thornton's article about the Waldron Creek Fire in the *Choteau Acantha*, I had no idea the role it would begin to play in my life going forward. As a smokejumper at the time, and one with a particular interest in fires where there had been fatalities, the story pushed all the right buttons for me. Just who were these five men who died, and what were the circumstances that led to their deaths? Why was so little known about this fire? The fact that the blaze occurred in an area where I had spent considerable time when I was growing up only added to the intrigue.

It's not just me. People in general seem to be fascinated with wildland fires, especially those that have led to fatalities. The Macleans, both Norman and his son John, weighed in with *Young Men and Fire* and *Fire on the Mountain*, respectively. In both of the fires they wrote about, a significant number of individuals perished. Norman's book detailed the Mann Gulch Fire of 1949 in Montana, where twelve smokejumpers and a recreation guard (ironically, a former smokejumper) died when they were caught in the cured-out grasses of a lonely Montana hillside. The book is a classic—my favorite, in fact.

John's book focused on the deaths of fourteen firefighters in 1994, all of whom succumbed when fire in the gambel oak of a Colorado mountainside raged uphill, overtaking them as they attempted to flee. The younger Maclean's book is as good as his father's, and I am proud to call him a friend. The infamous fires of 1910 likewise spawned a plethora of books, including Timothy Egan's *The Big Burn* and Stephen Pyne's *Year of the Fires*. In the fires of 1910, 87 people lost their lives, including 78 firefighters. Amazingly, this was far from the most deadly fire in history. That dubious honor belongs to the Peshtigo Fire in Wisconsin (1871), which burned so voraciously that an accurate count of the dead is impossible to come by. Some reports have put the loss of life at nearly 2,500 people. An account of this fire can be found in Peter Leschak's book *Ghosts of the Fireground.*

Why is it that we are so morbidly fascinated with death by fire? As humans, we seem to have a complex, almost contradictory relationship with fire—drawn to it and repulsed by it in seemingly equal measures. We love fire for the comforts it provides but simultaneously reserve some innate fear of it and what it might do if left unchecked. Part of this apprehension must come from wondering how we would respond if faced with the prospect of being burned alive. There are so many different ways to die, but this is not an option many would choose. I cannot help but think of those people in the World Trade Center on September 11, 2001, who, when faced with the choice of burning to death or committing suicide by jumping out the windows of the high towers, opted for the latter.

As a firefighter myself for almost twenty years, I never spent much time thinking about what it would be like to burn to death. That's just not somewhere I wanted my mind to go. Over the years, I inevitably had a few close calls. During my rookie season in 1989, my crew boss and I ended up responding to a one-thousand-acre fire all by ourselves. Multiple other incidents were keeping the rest of our crew busy, so he and I were the only ones available to attack this particular blaze. We did what we could, but it hooked around us, necessitating a hasty retreat on our part. Foolishly, we had parked the fire engine in an area of unburned grasses and light fuels. That spot became fully engulfed when the flames outflanked us, and two of the tires on the truck caught fire. Hastily, with Goodyears still ablaze, I moved the rig out of the conflagration and into an already burned area and then sprayed water on the tires to extinguish the flames.

I lost my hardhat in the scramble to get the engine out of harm's way. Once the fire had cooled, I went back to look for it. I found it—charred and melted to the point that it was nearly unrecognizable. I kept the remains on my shelf for years after that. The irony was lost on me at the time. I didn't want to think about what it would be like to be trapped by fire, yet I proudly displayed a piece of my own equipment that had suffered that very fate.

As noted earlier, when I read Ms. Thornton's *Acantha* article in 2003, it created an internal spark within me. I decided then and there to learn as much as I could about the fire and, if possible, write a book about it.

At times, that initial ember flared up, and I was able to research various aspects of the fire and write about them, but for the most part the book idea smoldered, continuing to smoke in my mind but showing little growth on actual paper. This process continued for the next decade. Through major life changes, such as the births of my two children and an occupational conversion in 2006 from firefighter to university professor, as well as through the more mundane elements of everyday existence, my interest in writing a book about the Waldron Creek Fire persisted, but production remained frustratingly low.

A goal is kind of like farting. It's a productive start, but until you actually take a dump you haven't produced anything tangible yet. That is the way I felt about trying to write this book. The idea germinated in my head for some time, but the actual word count remained pretty meager for years. I kidded myself that it boiled down to some love/hate relationship I have with writing. I loved the final product, but the process of getting there just absolutely pained me to no end. I knew what I wanted to say, yet I struggled with finding the perfect words to turn those thoughts into tangible print. I approached writing as on par with getting a vasectomy or having a prostate exam (I have had both, so the analogy worked). There was always something better to do to occupy my time. I distracted myself in any way possible to avoid having to write. Eventually, through practice, I learned to focus less on perfection and more on simply getting some words typed, words that I could work on later after having some time to digest them. I got better at this process but never mastered it.

During my first several years here at the University of Montana (UM), I used the excuse of tenure to rationalize not working on the book. If you receive tenure, it basically means that, unless you do something crazily egregious, you are assured a job each year. While some criticize tenure as a "guaranteed contract for life," it simply ensures that once you have it, you cannot be terminated unjustly simply because you covered a contentious topic in the classroom or carried out some controversial line of research that the administration did not support.

At the UM, a tenure-track faculty member must receive tenure by the end of his or her seventh year. A person is eligible for tenure after five completed years of service, which means that you have a two-year window in which to be granted tenure. So a person can put in for tenure during his or her sixth year and, if it is not received, still have one more shot the following year to apply again. Or you can roll the dice and apply during your seventh year and hope tenure is attained. If you do not get tenured by the end of your seventh year, the university has the right to shit-can you, a privilege that it exercises rather freely. Of course, since it's higher education, it doesn't call it that, instead going for some long-winded euphemism like "non-renewal of contract due to tenure denial."

The UM, like most other colleges and universities, evaluates its tenure-track faculty members based on three areas: teaching, research and service. While doing all of them is vital, they are not necessarily equal in importance. It seems, more and more, that research (specifically bringing in research funding and grants) is what gets a person noticed and subsequently tenured. The old adage of "publish or perish" most certainly holds true as well. In other words, in addition to bringing in research dollars, the faculty member must also be actively creating, writing and publishing scholarly works—the more the merrier. The problem with writing a book is that it takes so long. It's hard to build up an extensive list of published works when all your energy is going into one project. One of my main concerns was that if I got "head down" into a book project, I might be sacrificing my chances of tenure.

Well, as luck would have it, I ended up getting tenured. I applied for it after my fifth year at the UM, when I first became eligible. I was somewhat surprised to find out I had received it. Since it takes several months after turning in your application to find out the results, mentally I had prepared myself for the possibility that I might not get it. If that had been the case, I still would have had one year left to address my deficiencies and apply the following year. Truthfully, though, I had begun to accept that if I did not get tenure, even after my final year of eligibility, it wouldn't be the end of the world. In fact, it just might be a blessing. It would be a message from my peers and superiors that perhaps I needed to look into another line of work. Besides, I could always go back to smokejumping full time. Some days, the thought of doing so was highly appealing.

Don't get me wrong, I absolutely love my university job, but like other jobs, there are parts of it that are less than engaging. At least for me, the system seems to have built-in mechanisms that are designed to make you feel inferior, like you are never doing enough. Despite the fact that I am constantly busy, working long hours and frequently taking work home with me, I often end up worrying about not being productive enough with my time. I have learned that this is how higher education works: in order to keep professors motivated, you are frequently being evaluated and reviewed, with the not so subtle threat that if you don't produce, your career is in jeopardy. My beautiful wife reminds me frequently that the corporate world, of which she is a part, works the same way. I understand this, but it doesn't necessarily make it any easier to accept.

Often, the biggest impediment to getting any work done on the book seemed to be my professor job itself. While it is expected that a professor should produce some sort of research or scholarly work, the time to do so is limited. That challenge is magnified early in one's professorial career because so much time needs to be invested into developing and preparing courses. At the beginning of each semester, I would try to convince myself that I would devote an hour each day to working on the book. Unfortunately, as things ramped up—which they did every semester with rising-tide predictability—that always ended up being the first hour that got sacrificed from my schedule.

Another solution that would have addressed this problem of not having enough time would have been learning to say one simple word, that word ironically enough being one of the shortest in our complex English language: *no*. No as in, "I would love to help, but *no* I can't because I have too many other things going on right now." That would have fixed a bunch of my time issues right there. I wish it could have been that easy.

I will illustrate my point. One morning a while back, just like pretty much all mornings, I checked my e-mail first thing when I got to the office. One of the e-mails I received was as follows, sent from a high-ranking university administrator. It was the kind of e-mail that has your full attention before you have even opened it, simply based on who sent it. I shall paraphrase its contents and change the names to protect the guilty:

> *Dr. Palmer,*
>
> *Several weeks ago, I asked Dean Wormer to nominate/recommend a faculty member to add to a small team of faculty that has been meeting since last summer to devise a "big questions" curriculum to provide students with an education for the global century. That group attended a summer institute sponsored by the American Academy of Colleges and Universities and proposed a curriculum that motivates students around big questions and provides a pathway through the general education requirements. We plan to pilot facets of this curriculum in the fall, starting with 5 freshmen seminars, a set of linked courses, and funding for future requirements for these students to participate in co-curricular activities (such as study abroad, internships, research, service-learning) and a culminating capstone project (perhaps a performance, a research project, a creative work).*

The faculty/administrators currently participating on the team include Eb Onee, Director of African-American Studies; Anne Teak from Sociology; Ima Bisi from Student Affairs; Molly Quell from Chemistry; Gene Poole from Biology; and me. We are hoping you will join us as we mount a pilot. I can answer questions you might have about what we are doing and send you information about the overall proposal.

Thanks for your attention.

Jerry Attrick, PhD
Associate Provost for Undergraduate Education
Professor of Gerontology

With my perverted sense of humor, I could not stop laughing after I read, "We are hoping you will join us as we mount a pilot." All right, let me get this straight. The six of them are going to develop some "big questions" curriculum, and then, for some reason, after that they plan on putting an aviator into a sexually compromising position and they're hoping I'll join the party? It's going to get pretty awkward if we are all supposed to do our mounting at the same time. God, my mind is polluted. I can make a sexual innuendo out of just about any comment. I was very close to replying:

Dear Dr. Attrick,

I will gladly assist you and your group as we collectively attempt to mount a pilot. My preference is towards flight attendants (tall, hot, female), but I can adapt.

Respectfully,
Charles G. Palmer, Ed.D.

OK, seriously, now would have been a great time to say the "n" word. Not *that* "n" word, the other one. No. As in, "No thanks, Dr. Attrick. I'd love to help, but I'm too busy with other things right now. Really, it looks like an intriguing endeavor, but I've got my own project I am trying to work on right now." But the academic reality of it is that when an administrator at Dr. Attrick's level asks you to take part in something like this and you don't yet have tenure, saying "no" is really not one of your options. So, of course, I agreed to the request, and the committee became an absolute time-sink. The project itself was very interesting and had some pretty far-reaching implications as far as undergraduate curricula are concerned—I would even go so far as to say I enjoyed working on the committee—nonetheless, every hour I spent with them was one hour that could have been used writing this book. Welcome to my life.

Initially, my work on the book was limited to summers and the break between fall and spring semesters. Even then, though, I had plenty of other obligations on my plate, so it was not like I was able to devote all this time to the book. I hoped that would change, but as the old saying goes, hope is not much of a plan. I did have one potential option, though.

My grand evil scheme entailed putting in for a sabbatical, which would provide me with a large chunk of time to spend writing, and hopefully finishing, the book. Sabbaticals, at face value, seem to be these awesomely oxymoronic ventures where the university pays you to be gone. Instead of having to teach classes, advise students, attend countless meetings and do almost all of your other duties, a sabbatical clears you from all of these obligations while still providing compensation. How cool is that?

Of course, there are some catches. The university does not grant sabbaticals so you can go fishing and snowboarding whenever you please (there is a program that allows these things, but it's called termination, and it does not pay very well). Sabbaticals are competitive. There are a limited number of them granted every year, and best as I could determine from researching them, everybody wants one. An interested faculty member has to submit a lengthy application, and the university expects that, since it will be paying you to be gone, you will be productive with your time while on sabbatical. A person granted leave of this type has to guarantee that he or she will not only be carrying out some important research but also that this research will reflect favorably on the institution. In other words, it's a classic game of "If you scratch our back, we'll scratch yours." Works for me.

I put in for a sabbatical during the fall semester of 2012. The applications were due on October 15, the same due date for everything else of importance on campus, it seems. I invested a fair amount of time into completing the request, and then I shipped it off to my departmental chairman. One other person in our fairly small department was applying as well. I did not see this as that big a deal, since two of our colleagues had received sabbaticals at the same time a couple years back. Throughout the remainder of the semester, I frequently fast-forwarded myself in time one year, wondering what it would be like to not have any university obligations other than to work on my book. Not a great idea, but I couldn't help myself.

In the middle of January 2013, during the break between fall and spring semesters, I bumped into our chair in the not very large copy/mail room. "Sorry to hear that you didn't get your sabbatical," he said to me. The look on my face must have tipped him off to the fact that something was amiss. Very quickly, he realized I had not yet heard this news. He had received a letter from the provost telling him that my sabbatical request had been denied, but this information had not yet reached me. My colleague in the department who applied did get approved, but I was rejected. "I'm sorry," the chair said. "I thought you already knew." I mumbled something in return, walked back down to my office, closed the door and sat down.

All fall I had been telling myself not to get too excited about the sabbatical, since it was in no way a guarantee. Still, when I got the news, I was bummed. But in the grand scheme of things, on a severity scale of one to ten, this was maybe a two. I later learned that of the forty-one people who had applied for a sabbatical, thirty were successful. It didn't help my wounded pride to know that three-fourths of those who applied for a sabbatical got one and I wasn't on that list. But over and over throughout my life, I have learned that everything seems to happen for a reason. I guess it just wasn't meant to be for that year. I decided to throw my hat into the ring the next year and see what happened.

Well, wouldn't you know, next year was upon me before I knew it. Since my application was denied the first time, I decided to meet with the dean of our College of Education and Human Sciences to get some feedback from her on what I could do differently this time. After all, the rejection letter I had received from the provost said I was to consult with her should I have any questions. Dean Roberta "Bobbie" Evans is a classy lady and a top-notch leader, so her opinion would carry a great deal of weight with me. Bobbie was frustrated that I had not received a sabbatical on my first attempt and was very interested in helping me find success the second time around. When I met with her, she arranged for one of the Sabbatical Committee members (which is composed of faculty members from across campus) to also be present to help answer some of my questions, to give me some insight into how the selection process works and, most importantly, to give me suggestions on how to improve my application. The Sabbatical Committee, I learned, reviews the applications and forwards a list of its choices to the provost, who makes the final determination of who is selected.

Both of these ladies suggested I needed to do a better job of highlighting my previous fire experience and that I needed to include examples of the work that had already been done on the book. That way, they told me, it would be easier for the selection committee and the provost to see that this book was really just an extension of my previous research efforts here at the UM. Once again, the deadline for sabbatical application was October 15, so throughout the fall semester of 2013, I thought about what needed to be included. Luckily, in January 2013, a very well-written article about some research I and others had conducted into the prevalence rate of attention deficit hyperactivity disorder (ADHD) in wildland firefighters had appeared in a widely circulated UM publication called *Vision.* I included this piece in my application, as well as a sample chapter from this book. Additionally, I added a story from the *Great Falls Tribune* that highlighted some of the work I was doing on the book. While finishing this book was again listed as my top priority for sabbatical, I also integrated some other goals I wanted to accomplish. This entailed me promising to continue pursuing funding for some other fire-related research, such as the aforementioned ADHD work, and two investigations into the impacts of wood smoke on the cognitive function of humans.

On the morning of January 9, 2014, I received a call from Dean Evans. "Have you checked your campus mailbox today?" she asked. I had not. She could not restrain the enthusiasm from seeping into her voice. "Well, I just received a letter from Provost Brown informing me that you got your sabbatical. Congratulations!" I was somewhat shocked. Reflexively, I had begun to prepare myself mentally for the possibility that I would not get it after my second attempt. I had even gotten to the point where I had rationalized not being awarded a sabbatical. "Screw it," I told myself. "If the university is not going to support me in this, I'll figure out a way to get it done without their help." That all changed after the dean's call.

One afternoon in late January 2014, I decided to take a walk from my office. I had lately started walking and riding my bike more frequently. It gave me not only a chance to exercise but also time to think some more about this project. Missoula has a fantastic trail system along the river, so that's where I headed. My journey brought me through the Old Sawmill District, which is a forty-six-acre former mill site on the south bank of the Clark Fork River. For nearly one hundred years, this place in the center of town had been the hub of the regional timber industry. Initially called Polley's Mill after founder Edgar Polley, it was renamed Intermountain Lumber in 1956. Logs were floated down both the Big Blackfoot and Clark Fork Rivers and processed here. Dormant for about the past twenty years, the site is now under redevelopment, with the goal of turning it into a mixed-use space, complete with residential, retail, commercial and recreational opportunities.

Polley's Mill, Missoula, Montana, in 1924. *University of Montana Archives and Special Collections.*

Polley's Mill, Missoula, Montana, in 1930. *University of Montana Archives and Special Collections.*

Polley's Mill, Missoula, Montana, date unknown. *University of Montana Archives and Special Collections.*

Part of the revitalization efforts included a fourteen-acre public space called Silver Park.

I found myself being drawn to this place over and over again. It was as if I were being pulled by an enigmatic force I could not quite understand. There was some reason why my walks and rides almost always included a stop here. For the longest time, the motive eluded me. Then one day, part of the way through writing this book, the proverbial smoke in my subconscious cleared, and I was able to finally understand just why this place was so powerful. More on this later.

Work on Silver Park was nearly complete, and on this waning day in January, two men from the landscaping company were putting some finishing touches on their labor. The park was filled with pleasing odors: freshly planted trees and shrubs and recently tilled soil. I struggled with whether to say anything to the men, but then, on a whim, I decided I would. "Is this your project?" I asked one of them.

He stared at me for a second, unsure of how to respond. "Yeah, it is," he finally replied, with a hint of wariness in his voice. "Well, it's absolutely beautiful. You guys have done some great work here," I told him. His face lit up, and he said, "That's why I do this job. For comments just like that. Thanks."

Old Sawmill District, Missoula, Montana, in May 2015. *Author's collection.*

Old Sawmill District, Missoula, Montana, in March 2015. *Author's collection.*

That, I believe, is what many of us are looking for in our work—and for that matter, our lives in general. The appreciation and recognition from others that what we do is meaningful, that it adds value to society in some way, no matter how big or small. The knowledge that we have made an impact. The five men who died in the Waldron Creek Fire never received that from the job that ended their lives. They made the ultimate sacrifice, yet their loss has been forgotten. This book is an attempt to tell their tale, what little is known of it, and to make others aware of their contributions to the greater good. It is the least that can be done for them, even if it is more than eighty years too late.

ACKNOWLEDGEMENTS

A book cannot be written in a vacuum. Many people helped me along the way in this endeavor, and I am grateful for their assistance. First and foremost, I would like to thank Nancy Thornton for making me aware of the Waldron Creek Fire and for all of her aid as I researched the event. This book would have never happened without her. Professionally, I am indebted to Provost Perry Brown, dean of the College of Education and Human Sciences Roberta Evans, Health and Human Performance (HHP) chairman Scott Richter and all of my colleagues in the HHP Department for providing me the time and support needed to complete this project. Thanks to Mark Fritch and the University of Montana's Archives and Special Collections for access to many of the photographs included, to Shawn Mihalik for his fine editing work, to Jim Heckenbach for his genealogical expertise and to Artie Crisp, Jaime Muehl and the other professionals at The History Press for their professionalism. Ottis Bryan, Joan Sharpe, Ken Robison, Mark Smith and Thomas Taylor also made valuable contributions to this project. Personally, a heartfelt thanks to my wife, Christine, and children, Summer and Skye, for putting up with me while Dad "worked on his book." And lastly, cheers to my parents, Doug and Marilyn Palmer, for without them I would not be who I am today.

PROLOGUE

August 25, 1931, approximately 5:00 p.m., South Fork of Waldron Creek, thirty-five miles west of Choteau, Montana.

All twenty of the men looked about with growing apprehension. Some had even begun to pray, but they did so quietly and under their breaths, not wanting to appear weak or afraid in front of the others. Just moments earlier, when their guide had left them unattended at this small spot fire on the Rocky Mountain front, things had been relatively calm. Now the wind speed had ratcheted precariously upward, and the fire—which had up until this point been burning rather innocuously on the ground—was climbing the ladder fuels[1] and beginning to torch large pockets of trees.

A voice pierced the smoke and confusion: "Follow me. We can get this thing!" Later reports would suggest that the command came from Hjalmer "Harry" Gunnarson, thirty-nine, a Canadian and a veteran of the Great War. However, with the freight train–like rumble of the fast-approaching blaze and the rising panic among the group of twenty fire recruits, no one could be certain who had spoken these words or, for that matter, exactly just what had been said. Four other men picked up their tools, perhaps in answer to the challenge: twenty-year-old Herbert Novotny and twenty-four-year-old Frank Williamson, best friends from Great Falls; forty-six-year-old drifter Ted Bierchen, originally from Luxembourg; and Charles Allen, a roughly thirty-seven-year-old from Pittsburgh, Pennsylvania. Now five strong, they moved as a small pack away from the rest. Lacking any

A Montana fire, date unknown. *University of Montana Archives and Special Collections.*

other leadership, the remaining fifteen men froze in place, paralyzed by a primal fear.

Later, some of those who survived the blowup would claim that this quintet of men panicked and attempted to outrun the mushrooming forest fire by traversing their way to the bottom of the gulch, only to be cut off when the blaze roared through the heavy timber located there. However, when investigators finally located their charred bodies, they found four of the men huddled within feet of one another, near what appeared to be a freshly dug trench and portions of constructed fireline.[2] (The fifth man, whose remains would not be found for another twenty-four hours, was located roughly seventy-five yards farther up the north-facing slope. It is still unknown why he was not with the others.) If they had, in fact, attempted to build a fireline, it would suggest that the men had not "cut and run" like some of the survivors indicated but had decided to go on the offensive against the blaze and had paid for that decision with their lives. If the other fifteen men had helped them, would the outcome have been different? Would the fire have been stopped in its tracks, or would there instead be twenty dead? We will never have answers to these questions, only hypotheticals.

Eventually, the fifteen other recruits, either through keen decision-making or blind-ass luck, were able to find their way to safety by running in the other direction, away from the quickly encroaching flames. Within days, three of these men would sign an "official" statement, which declared that "all possible precautions were taken by those in charge of the fire to provide for our safety." The county coroner, the man now responsible for dealing with the five dead bodies, would write on each man's official death record: "Came to his death by being traped [*sic*] in a forest fire. No one to blame but himself."

Hastily arranged funeral services were held, a couple short articles about the deaths appeared in local newspapers and then that was it. The deadly fire season of 1931 ended soon thereafter when heavy rains finally cooled the region. The Forest Service—and the world in general, for that matter—moved on. The men who died were expendable casualties in the Forest Service's war on fire. Questions were never raised. Investigations were never launched. The account of the Waldron Creek blaze and the five men who died there was closed, and for the next seventy-two years it remained that way. But in the June 11, 2003 edition of the *Choteau Acantha*, a story about the fire appeared. Written by local reporter Nancy Thornton, it detailed the accounts of the blaze and the men who were killed there. This is where the story picks up again.

1
THE BUILDUP

The year 1931. It was a fascinating time in American history—indeed, in the history of the rest of the world as well. By August of that year, when the Waldron Creek Fire claimed the lives of the "Forgotten Five," the Great Depression had been in full effect for nearly twenty-two months. While scholars continue to debate whether the collapse of the American stock market on October 29, 1929, was the cause of the Depression or merely a symptom, the end result was the same: millions and millions of individuals in the United States and around the world were out of work. Unemployment in the United States rose to 25 percent. In other countries, up to one-third of all workers were without jobs. International trade declined by more than 50 percent.[3] Economists like John Maynard Keynes and Milton Friedman would later speculate about why the Great Depression occurred, but that is all they offered: theories. The cold reality was that people's lives around the world had been turned upside down. For many, they would never right themselves again.

Herbert Clark Hoover, the thirty-first president of the United States, served in this nation's highest political position from 1929 to 1933. He had been in office only eight months when the stock market plunged on Black Tuesday. A Republican, Hoover was raised in a Quaker family and worked as a mining engineer prior to venturing into public service. As the head of the U.S. Food Administration during World War I, he gained a level of notoriety for his humanitarian relief efforts in Belgium. Later, in the 1920s, as the United States secretary of commerce, he helped to foster

"Negroes employed as drillers on the construction of the Hoover Dam." October 1932. *National Archives and Records Administration, ARC Identifier 293747.*

economic partnerships between businesses and government. Hoover was a strong proponent of the efficiency movement, which sought to identify and eliminate waste in all areas of society and the economy and to devise and implement best practices for achieving said efficiency.

As president, Hoover donated all of his salary to charity, a presidential act matched only by John F. Kennedy three decades later. When the Great Depression struck, Hoover tried to stimulate economic development through modest public works projects, such as the Hoover Dam on the Colorado River. Construction on the dam began in 1931, the same year as the fire, and after five years and the loss of some one hundred workers' lives, the project was at last completed. Despite his efforts, the economy continued to plunge. Historians generally agree that the worsening economic picture, coupled with Hoover's staunch support of the Prohibition movement, contributed to his loss in the 1932 presidential election to Franklin Delano Roosevelt.

As noted above, Prohibition was still in full effect in 1931. The Senate proposed the Eighteenth Amendment in December 1917, and after a lengthy political process, the United States went officially dry on January 17, 1920. Although alcohol was now illegal, speakeasies (also called blind pigs) and rampant bootlegging flourished as the demand for booze remained high. The ban on alcohol lasted until March 1933, when FDR signed into law the Cullen-Harrison Act, which legalized beer with a 3.2 percent alcohol content. Ratification of the Twenty-first Amendment in December 1933 repealed the Eighteenth Amendment, finally lifting the nation's embargo on alcohol.

The year 1931 also brought with it other developments and events, some with long-lasting implications and others that simply became footnotes in history. A gas explosion at a coal mine in Manchuria killed three thousand people. An earthquake in Nicaragua conveyed death to two thousand more. The State of Nevada legalized gambling. Construction on the Empire State Building—the tallest building in the world at that time, rising 1,454 feet into the New York skyline—was completed. Alka Seltzer was introduced to the American public. The original *Dracula* movie, starring Bela Lugosi, was released. John McGraw, baseball player and longtime manager of the New York Giants, commented that nighttime baseball games would never catch on with fans. Al Capone was finally indicted on charges of tax evasion and various violations related to Prohibition. And on August 21, Babe Ruth hit his 600th homerun off of George Blaeholder of the St. Louis Browns.

The Duval children and their dog, Montana, in 1930. *University of Montana Archives and Special Collections.*

The Boehme-Cummings Trout Fly Factory, 12C West Main Street, Missoula, Montana, in 1931. *University of Montana Archives and Special Collections.*

Opposite, top: The interior of a wolfer's cabin. Montana, 1902. *L.A. Huffman. University of Montana Archives and Special Collections.*

Opposite, bottom: Main Street in Miles City, Montana, 1880. *L.A. Huffman. University of Montana Archives and Special Collections.*

Interesting happenings were occurring in Montana, as well, during 1931. The Knights of the Ku Klux Klan chapter was still active in Butte, although its numbers had dropped significantly from the height of its influence in the mid-1920s. While outwardly it pronounced its enemies as corruption and vice, the Klan's real targets were Catholics, blacks, Jews and those born in foreign lands. Butte, teeming with immigrant Irish Catholic miners, was fertile ground for the KKK to sow its seeds of hatred. Membership qualifications consisted of being "native born, white, Protestant, Gentile, and an American citizen."

The western meadowlark was adopted as the official state bird in March 1931. Work began on the Beartooth Highway, linking Red Lodge to Yellowstone National Park. Five years later, at a cost of $1.1 million, that project was finally concluded. L.A. Huffman, photographer/school board member/county commissioner/Montana House of Representatives member from Miles City, died that year as well. His photographs, primarily focusing on Plains Indians and cowboy ranchers, are considered to be some of the finest taken during this period.

INTERIOR OF A WOLFERS CABIN MONTANA 1902

And in 1931, at the behest of a schoolteacher by the name of Maureen Hayes, Mike Mansfield applied to what was then Montana State University in Missoula (later named the University of Montana) in spite of the fact that he had dropped out of school before finishing the eighth grade. He was provisionally accepted, with the stipulation that he finish high school equivalency classes before enrolling, which he did. Mansfield had left school at the age of fourteen, lied about his age and enlisted in the U.S. Navy so that he could fight in World War I. He completed several convoy missions across the Atlantic before the navy discovered his real age and dismissed him from service. Promptly thereafter, Mansfield joined the army and then went

A mule train on the 1,100-foot level, Rarus Mine, Butte, Montana, circa 1910. *University of Montana Archives and Special Collections.*

Maude Evelyn Lehsou
Missoula
French
Alpha Phi

Fae Logan
Deer Lodge
Business Administration
Delta Delta Delta

Hugh Joseph Lemire
Ronan
Law
Sigma Chi

Kenneth Eckford Lord
Great Falls
Business Administration
Sigma Phi Epsilon

Margaret S. Lewis
Roundup
Foreign Languages
Alpha Delta Pi

Melvin E. Magnuson
Helena
Business Administration

Rubin Lewon
Glasgow
Chemistry
Kappa Sigma

Michael Joseph Mansfield
Great Falls
History

James Thomas Likes
Missoula
History

Don Franklin Marrs
Harlowton
History

Glenn Eugene Lloyd
Great Falls
Business Administration

Mary Elizabeth Martin
Anaconda
Biology
Kappa Delta

Mike Mansfield as a senior in the *Sentinel* yearbook, Montana State University, Missoula, Montana, 1933. *Author's collection.*

into the U.S. Marine Corps. With the marines, he visited the Philippines, Japan and China, which sparked in the young private a lifelong interest in the Asian regions.

Mansfield was honorably discharged from the marines in 1922. He returned to Montana and for the next eight years worked as a mucker,[4] shoveling ore a half mile below ground in the copper mines of Butte. With the financial help and moral support of Ms. Hayes, whom Mansfield would later marry, he enrolled at the UM. He would go on to earn bachelor of arts and master of arts degrees and was awarded a faculty position at the UM teaching Far Eastern and Latin American history and political science. Mansfield also began work on a PhD at the University of California–Los Angeles. These plans changed when, again at the urging of his wife, he decided to enter politics.

After an unsuccessful bid for the House of Representatives seat in Montana's first congressional district in 1940, Mansfield ran again in 1942, easily defeating Republican businessman Howard Hazelbaker. He succeeded Jeanette Rankin, who had decided not to seek reelection. Rankin holds the honor of being the first woman elected to Congress (1916) and also is known for her opposition to the United States' entry into both world wars. After serving five terms in the House, Mansfield opted to run for the Senate against incumbent Zales Ecton. Senator Joseph McCarthy of Wisconsin—he of the infamous McCarthyism anti-communist movement—came to Montana to campaign for Ecton.

During this campaigning, McCarthy chose to publicly question the patriotism of candidate Mansfield. It was an astonishingly bold tactic considering that Mansfield had fought for his country in the Great War and had served honorably in three different branches of the military. After scraping out a narrow victory, Mansfield arrived in the Senate, only to be greeted by Senator McCarthy asking him how things were back in Montana. "Much better since you left, sir," was Mansfield's classic retort.

In 1961, Mansfield was unanimously elected Senate majority leader, a position he held for sixteen years, until his retirement in 1977. This made him the longest-serving majority leader in the history of the Senate. After leaving the Senate, he was appointed to serve as the U.S. ambassador to Japan, a post he held until 1988. After his retirement in 1989, he was bestowed with the Presidential Medal of Freedom, the highest civilian award in the United States. Ten years earlier, in 1979, the University of Montana had named its new campus library the Maureen and Mike Mansfield Library. Much of the book you are now reading was researched and written in this very building.

Senator Mike Mansfield riding in a car with President John F. Kennedy, taken outside the Mansfield home in Great Falls, Montana. September 26, 1963. *University of Montana Archives and Special Collections.*

Meanwhile, in Choteau, Montana, the settlement closest to where the Waldron Creek Fire would eventually burn, life in August 1931 was moving along at the typical small-town pace. The annual flower show, sponsored by the Choteau Women's Club, fielded a record 136 entries. This far surpassed the previous record of 105. Rodeo promoters were working on a two-day event, which was to be held on September 6 and 7. Thirty riders were expected and twice that many horses, twenty-seven of which had never been ridden before. Due to the drought, the upland bird-hunting season statewide was cancelled, and the duck- and goose-hunting seasons were going to be postponed from September 15 to October 1.

By far, the talk of the town was focused on the city marshal, A.H. McGinnis, and the near fatal beating he had received at the hands of a robber. Early on the morning of August 22, the Choteau Hardware Company had been burglarized. The thief made off with about $300 worth of guns and ammunition and a large amount of money from the cash register. Later that day, a local man discovered a cache of stolen weapons near the Great

Northern Railroad tracks, a short distance south of the J.M. Baker grain elevator. He alerted McGinnis to the find, and McGinnis staked out the spot that evening, waiting for the robber to return and claim his loot. Shortly after he had begun his surveillance, an individual approached the guns. McGinnis ordered him to stop. While trying to take the man into custody, a terrific fight ensued. McGinnis discharged his .32 automatic to warn the suspect, but the gun jammed because he had used the wrong caliber of ammunition.

According to the newspaper report from the *Choteau Acantha*, the thief "beat, kicked, and choked the officer, breaking one or more of his ribs and cutting gashes to the bone in his head and face." The assailant subsequently got away, and McGinnis finally succeeded in getting himself to the Choteau hospital, "exhausted from loss of blood and other injuries." Upon hearing of these events, a large number of locals organized a manhunt. Eventually, they found the thief's camp and more of the missing guns, but they never were able to capture the crook. While these events fed the local gossip mill for the next several days, the topic of conversation was about to make a sudden change. Soon, five men would be dead, burned to death in the mountains just outside of town.

Waldron Creek

When it comes to the history of the Waldron Creek Fire, the complete story of what happened that day actually begins much earlier than August 25, 1931, the day the five men died. Actions and inactions days, months and even years prior to this date contributed, in both large and small ways, to their deaths. These factors cannot be ignored. Failing to give them the attention they deserve results in an incomplete understanding of the upstream tributaries that factored into the demise of not only these five individuals in Waldron Creek but also dozens of other firefighters in years prior to and after 1931. Therefore, in order to better understand what happened on August 25, 1931, it is important to place this specific event into a broader historical context.

The Early Forest Service

When the Waldron Creek Fire occurred in 1931, the United States Forest Service was still a relatively nascent organization, complete with the growing pains and immaturities experienced by all new entities, especially those of the governmental variety. In 1905, the signing of the Transfer Act by President Teddy Roosevelt shifted the management of American forest reserves from the General Land Office of the Interior Department to the Bureau of Forestry, which would become known as the U.S. Forest Service. Under the act, more than sixty-three million acres of forest reserves and five hundred employees were transferred to the new outfit.[5] Almost overnight, the Forest Service became a major player in American government simply because of the landmass it controlled. Gifford Pinchot, a Yale forester, was appointed its first chief forester.

This newly created Forest Service found itself caught between the dynamic tension of two diametrically opposed groups: large timber companies that saw the nation's forest reserves as a crop to be harvested and the fledgling forest preservationist movement, which was strongly resistant to any commercialization of the woods. Trained as a forester, Pinchot was interested in how the principles of scientific management could be applied to the long-term profitability of our nation's forests and natural resources. As such, his philosophies contrasted sharply with those of the large, powerful timber companies, which were much more interested in short-term gains via massive clearcutting[6] of trees. Likewise, Pinchot's beliefs brought him into opposition with the likes of John Muir and other preservationists, many of whom saw the forests as being off limits to any and all development.

With the election of William Howard Taft as president in 1908, Pinchot's days as chief forester became numbered. A series of policy disputes between Pinchot and newly appointed secretary of the interior Richard Ballinger led Taft to dismiss the chief forester in early 1910 and replace him with another Yale graduate, Henry Solon Graves. It would not be long—mere months, actually—before the Forest Service and its new leader would face their greatest challenge.

The Fires of 1910

In the span of two days, August 20 and 21, in the summer of 1910, hurricane-force winds enabled hundreds of small fires in the Northern Rockies to combine into one "Big Blowup." In its wake, this massive firestorm killed eighty-five people, mostly firefighters; scorched over three million acres in Idaho and Montana; and left entire towns and forests in ashes. The fires of 1910 would exert a tremendous influence on fire policy. Chief Graves, wanting at all costs to avoid a similar scenario in the future, laid out a platform for the Forest Service that involved defeating wildfire and removing it from the landscape if at all possible.

This emphasis on excluding wildfire from the natural environment contributed to the creation of the 10:00 a.m. Policy, a Forest Service stratagem that called for gaining the upper hand on all fires within a short period of time. This rule of engagement was still very much in effect when the Waldron Creek Fire burned. For evidence of this, one need look no further than the 1931 "Report of the Forester," written by Chief Robert Y. Stuart and submitted to the secretary of agriculture and Congress as a whole. Stuart wrote, "Effort is being focused on the objective of control of every fire before 10 o'clock of the morning after its discovery, regardless of how far or fast it may run initially."[7]

The fires of 1910 also provided the Forest Service with an early opportunity to perfect an act that it would repeat with the victims of the

The railroad tracks and burn area in Saint Joe Forest, Idaho, circa 1910. *University of Montana Archives and Special Collections.*

The burned-over lodgepole area above Seeley Lake, Montana, circa 1910. *University of Montana Archives and Special Collections.*

Waldron Creek blaze and many other fires: reluctance to acknowledge or fully address its accountability in the deaths of those who perished in the line of duty fighting its blazes. Val Nicholson was a seventeen-year-old firefighter who died after falling into the Bullion Mine while battling one of the 1910 fires. For some unknown reason, the task of retrieving the body fell to his father, who navigated the still-burning terrain and dragged the body down to Wallace, Idaho, for burial. Within two weeks, the grief-stricken father was overwhelmed by his own anguish and died of a heart attack. Mrs. Nicholson, now widowed with young children to feed, pleaded with the Forest Service for some form of compensation. No can do, the government replied. She then simplified her request: a headstone for her son's grave. Forest Service officials told her they would check into it. She never heard back.[8]

However, the treatment of one of its own, Edward Pulaski, might provide the greatest evidence that the early Forest Service had little interest in recognizing or caring for those who were injured or died fighting its fires. Ranger Pulaski, along with forty-five other men, was on a fire about ten miles southwest of Wallace, Idaho, when the Big Blowup hit. Barely outrunning the conflagration, Pulaski led his men into a small mineshaft. He ordered them deeper into the mine. Pulaski stayed at the mouth of tunnel, hanging wet blankets over the entrance and threatening to shoot any man who attempted to escape. Despite

Edward Pulaski. *Photo courtesy of USDA Forest Service.*

suffering significant burns to his body, particularly his eyes, Ranger Pulaski was able to save the bulk of the crew.

Scarred by the fires of 1910, both internally and externally, Pulaski sought relief. In recompense for his physical injuries, he asked the Forest Service to help cover some of his medical expenses and lost wages. Again, the answer was no. For his posttraumatic stress–related issues, he turned

Right: The Pulaski mineshaft, circa 1910. *Photo courtesy of USDA Forest Service.*

Below: An unknown woman standing at the mouth of the Pulaski tunnel, outside Wallace, Idaho. *University of Montana Archives and Special Collections.*

inward. An inventive type by nature, he found that working with his hands was an effective therapy for calming his troubled mind. By 1913, after much trial and error, Pulaski had developed the penultimate firefighting tool: a combination axe on one side and mattock hoe on the other. Despite the facts that the tool now bears his name and today the General Services Administration puts out bids for more than thirty-five thousand new Pulaskis each year,[9] Pulaski himself never made a dime from this creation.[10]

Pulaski also took it upon himself to care for some of the remains of those who were killed in the 1910 fires. As noted by Timothy Egan:

> *Six years after the fire, Pulaski still tended the graves, still struggled to get through the day without pain, still lobbied for a proper memorial for the dead. He wrote a memo to the government outlining the costs of his proposal: Concrete, $200. Granite slabs engraved with the names of the dead, $215. Soil and grass, $20. But for his $435, the Forest Service would need an act of Congress, he was told.*
>
> *Eleven years after the fire, in 1921, Congress appropriated $500 to the Coeur d'Alene National Forest "for the markings of the graves in Wallace, Idaho." Pulaski used the money to get the names of the dead etched in stone. But other bodies were still scattered, and it bothered him. Finally, in 1933—nearly a quarter century after the Big Burn—a central graveyard was established on a grassy slope in the dreamy little town of St. Maries, Idaho. Remains were dug up from different locations and moved to the burial ground, at rest in one place at last.*[11]

The graveyard in St. Maries is named Woodlawn Cemetery. The Firefighters' Circle in Woodlawn comprises the graves of fifty-seven firefighters who died in the 1910 fires and on fires in the region during later years. Unfortunately, Ranger Pulaski was not alive to see the day in 1933 when the Firefighters' Circle was unveiled. He died two years earlier, in 1931, the same year the Waldron Creek Fire burned.

Drought and the Great Depression

Beginning in 1917 and continuing jaggedly for the next two decades, Montana experienced a drought of historical proportions. Over a six-year period, from 1919 to 1925, roughly two million acres of farmland moved out of production, and eleven thousand family farms, approximately one-fifth of the state's total, were abandoned.[12] Besides leading to a mass exodus of Montana's rural population, the desiccated grasslands and forests of the state were primed to burn. For those who stayed, jobs were as scarce as rainfall. Unfortunately, things were about to get even worse.

As detailed at the beginning of this chapter, the Great Depression descended on the United States in October 1929. In Montana, which was already struggling from ten-plus years of drought, the situation deteriorated even further. The primary sector of the economy was particularly hard hit. Industries that made direct use of natural resources—such as logging, agriculture and mining—felt the biggest blow. Since the economic wellbeing of the West was so closely tied to these trades, this region suffered significantly, as did the men who relied on these types of jobs for their financial security. The unemployment rate, which hovered at about 25 percent nationally, was even higher in Montana.

In review, when the fire season of 1931 arrived in the Northern Rockies, a diverse array of factors was firmly in place: the Forest Service was still an immature organization with tunnel vision focused on protecting its trees and putting out fires quickly regardless of conditions, and it had a poor track record of taking care of those who were hurt or killed fighting its fires; severe drought conditions had been in effect for more than a decade; and thousands of able-bodied men were out of work and looking for employment due to the Great Depression. Although relatively inert when examined individually, when added together and subjected to a considerable amount of heat, these factors combined to form a lethal blend. The fire season of 1931—when twenty men were killed fighting Forest Service fires, including the five men who died in Waldron Creek—would prove to be the deadliest for firefighters since the catastrophic losses in 1910.

Fire Season 1931 Leading up to Waldron Creek

Typically, the height of the fire season in the Northern Rockies region occurs around midsummer.[13] It is at this time that temperatures and winds are usually high, humidity and fuel moistures both low and storms with lightning frequent. The fire season of 1931 decided very early on that it was going to be anything but a typical year.

By April and May in the states of Oregon, Washington, Idaho and Montana, the combined factors of long-term drought, low humidity, wind and afternoon lightning storms had produced a fire danger that had never before been encountered at that time of year. Crews working on road construction projects for the Forest Service, who normally found moist soil at depths of one to two feet, were instead caked in dust even when digging at depths of four feet or more. Fires, usually infrequent and small at this time of year, instead were numerous and substantial in size. By the middle of July, when the usual summer drying pattern was added to the cumulative effect of long-term drought, the forests of the West had arrived at what the chief of the Forest Service would later call a "powder-keg state."[14]

On August 2, at approximately 11:00 a.m., the Priest River Fire in northern Idaho flared to life just outside the forest boundary. Pushed by strong winds, it crowned[15] immediately and raced through the tree canopies. One hour later, it was roughly 1,300 acres in size. By early afternoon, the fire had chewed through nearly 50,000 acres, with a flaming front five miles wide and a total length of fifteen miles. When nightfall finally came and the winds mercifully ceased, the fire had destroyed thirty-four ranches, completely blocked two major highways and trapped 100 people. The Forest Service quickly mobilized its forces, and by midnight, 700 men were fighting the blaze. Twelve hours later, that number would more than double, with 1,500 firefighters on the line.

A little over two weeks later, on August 17, more than sixty new fires were started in western Montana and northern Idaho when a "dry" electric storm swept over the area. Less than a week after that, a storm on the evening of Saturday, August 22, ignited twenty-eight more. Many of these fires escaped initial containment efforts and would eventually cause significant problems. As the lightning bust made its way over the continental divide, it sparked a number of other blazes on the east side of the mountains, including a small fire with a fatal future about thirty-five miles west of Choteau, Montana—in Waldron Creek.

2

THE FIRE

The Waldron Creek Fire started as a rather innocuous little blaze high in the headwaters of the South Fork of Waldron Creek. On Sunday, August 23, smokechasers Freeman Page and C.E. Spires first noticed the smoke from their vantage point on Mount Wright, roughly fourteen miles away in a northwesterly direction. Their job was to detect such smoke and, if necessary, do the best job they could of putting out the fires. Those were tall orders considering that, at the time, the Teton District of the Lewis and Clark National Forest in Montana consisted of an area of nearly a half million acres, or roughly two-thirds the size of Rhode Island, the state we love to pick on when making geographic comparisons. Getting to the fire was not easy, either. The *Choteau Acantha* later noted that access to the blaze was difficult, and it was "on a very bad place where it is hard to fight."

Complicating matters was the size of the available workforce to actually fight the fire or, in this case, the lack of such personnel. There were only three smokechasers in the Teton District, to go along with four fireguards[16] and another roughly ten men who did trail work and were available to fight fires if necessary. Many of these men were already committed to other fires or were simply unavailable. Page and Spires telephoned their bosses to brief them on the situation and then nervously kept their eyes on the growing blaze.

News of a fire in Waldron Creek worried Teton District forest ranger Walt Streeter. Born in Wisconsin in 1889 and educated there, Streeter and his family had moved to Choteau during his teenage years, so he was a hometown boy, so to speak. He was a veteran of the U.S. Army and had served in France during the Great War. He moved back to Choteau after World War I ended, married a local gal and went to work for the Forest Service as a ranger, a post he held until retiring in 1951. As a local, he knew the topography and the fuels in the area and was keenly aware that fourteen years earlier, in 1917, a fire had burned nearly 1,500 acres in the nearby vicinity.

The weather was not helping, either. So far in his district, 1931 was shaping up to be the warmest year ever recorded to date, and the total precipitation for the year was nearly 40 percent lower than average. Streeter knew that a fire in this area would be difficult to stop once it gained any momentum due to the thick stands of mature timber located throughout the drainage area and the fact that other incidents in the region were tying up available resources and manpower. The district ranger had another concern as well: much of the timber in the Waldron Creek drainage had recently been sold to a local logger, and it fell on Streeter's shoulders to keep it protected until it was harvested.

Forest Service rangers were a unique and special breed, and Streeter was no exception. According to recruitment posters of that era, rangers were expected to "build trails, ride all day and night, pack, shoot, and fight fire without losing their heads."[17] The Ranger Examination, three hours long and an exam that all aspiring rangers needed to pass with a high score to even be considered for employment, tested an applicant's knowledge on topics as diverse as cattle, horses, fire, forestry, wood, lumbering, grazing and land surveying. Once employed, rangers needed these skills, and a host of others, to effectively do their jobs. As noted by one historian, "The Ranger in his district was often the only policeman, fish and game warden, coroner, disaster rescuer, and doctor. He settled disputes between cattle and sheepmen, organized and led firefighting crews, built roads and trails, negotiated grazing and timber sales contracts, carried out reforestation and disease control projects, and ran surveys."[18]

One of the reasons that so much emphasis was placed on rangers having a strong background in fire was because, at this time, the Forest Service had no large, organized fire crews available. Those districts that contended with fires employed some smokechasers and fireguards, but only in small numbers. Instead, they relied on a "fire militia" strategy. Since it was so thinly staffed with full-time firefighting employees, the Forest Service simply hired temporary help (fire militia) when needed. Recruitment took place wherever one would expect to find men looking for work, with bars, pool halls, soup kitchens and employment offices being the most likely venues.

Major Kelly and Clyde Fickes, U.S. Forest Service Rangers, sit horseback discussing a forest fire. Date unknown. *University of Montana Archives and Special Collections.*

Since the Great Depression was in full swing, there were plenty of able bodies to choose from and even more who weren't prepared for the rigors of the fireline. The men were expected to furnish their own clothing and footwear. The government supplied the tools and the transportation; however, it did not provide any formalized training in how to actually fight fires. It simply hoped that the men who signed on already possessed these skills or that they survived long enough on the fire to learn them. It fell to the rangers to facilitate this recruitment process and lead those temporary firefighters who were hired.

On Sunday night, August 23, Streeter left from Great Falls en route to the fire, accompanied by fifteen men he had been able to recruit on short notice. On Monday, after getting as close to the fire as they could by vehicle, the men took off on a trail that snaked its way toward the South Fork. After several miles, the ragtag group left the main trail for a five-mile cross-country hike through downed timber and difficult terrain. By the time they arrived at the blaze, it had grown to nearly twenty acres in size, due primarily to heavy winds the previous night. While most of the fire was confined to burning on the ground, it would occasionally torch out single trees, quickly turning what had been robust green pines into smoking, blackened skeletons within seconds. This torching act unnerved many of the new recruits, who were unaccustomed to the raw power of fire. The men did what they could to check the growth of the fire, but inexperience and exhaustion from the grueling hike limited their efforts. It became painfully obvious to Streeter that he was going to need significantly more help.

Realizing the potential of the fire, Streeter turned command of the blaze over to H.H. Hendron, assistant supervisor of the Helena National Forest, who had also arrived at the fire earlier that day. Streeter retraced his steps and hurriedly returned to Choteau looking for more firefighters. In short order, he was able to gather about twenty-five additional men and more equipment and was soon back on the road to Waldron Creek. Before leaving, Streeter placed a call to the forest headquarters in Great Falls, telling its staff that if he was going to have any chance at all of catching the blaze, he was going to need many more men and resources to support the effort.

In Great Falls, the wheels of motion turned rather rapidly. Although other fires in the region had significantly tapped into the number of available men, there were still plenty left to choose from. Well before dawn on Tuesday morning, August 25, twenty additional men had been conscripted into service, having been hastily rounded up from the local pool halls, soup kitchens and bars around town. Little did the men realize that by suppertime of that same evening, a quarter of them would be dead, their bodies so grossly burned that those looking for them would have a difficult time distinguishing the remains from the blackened tree stumps of the landscape.

The twenty new fire recruits were loaded onto train cars for the roughly sixty-mile trip to Choteau. Once there, they packed into cars and trucks, circus clown style, for the twenty-plus-mile drive over the dusty, washboard road to the mouth of Teton Canyon. A makeshift fire camp was beginning to form here, complete with a camp cook and a supply depot of tools and equipment the men would need to fight the blaze. From camp, it was another twelve miles—all on foot—to get to the fire ground itself. The job of leading the men to the fire landed on the broad shoulders of a local farmer, Franklin Fellers. Fellers was an experienced woodsman and firefighter, and he knew the area well.

The long hike took its toll on the men, few of whom were conditioned to the demands the day had so far entailed. Nearly five hours after leaving the main camp, the group arrived at the edge of a twenty-acre spot fire about a quarter of a mile east of the main blaze. The men collapsed to the ground

Men working on a fire, date unknown. *University of Montana Archives and Special Collections.*

and gulped the lukewarm water from their canteens. Fellers left them and continued on to the main fire, where he checked in with the Teton District alternate forest ranger Bud McNeal. McNeal was in charge of the men already on the fire. Privately, the two shared their thoughts and concerns with each other. Both reached the consensus that their chances of stopping the blaze with the resources they presently had were not good, especially if the devilish afternoon winds picked up once again.

As if on cue, the breeze stiffened. Tendrils of flame, which had been confined to the ground, now jumped weightlessly into the tree branches and roared through the crowns. At this point, according to later reports, the main fire was more than four hundred yards away from the men, hidden behind a small ridge and on the opposite side of the gulch from where they now hunkered. Within minutes, the main fire would more than double in size, growing from three hundred acres to more than seven hundred. The two fires, main and spot, pulled toward each other in an effort to become one, with the men trapped in between. The sound was what they noticed first. It came across as a wildly misplaced train, barreling down on them from invisible tracks, belching black smoke through the trees.

The twenty firefighters were still without supervision. Fellers had not returned from his rendezvous on the main fire with alternate ranger McNeal. As the conflagration built itself up, communication among the men was breaking down. Testimony from some of the survivors would later indicate that Harry Gunnarson stepped up to fill the leadership void and that four others decided to follow his command: Herbert Novotny, Frank Williamson, Ted Bierchen and Charles Allen.

Hjalmer "Harry" Gunnarson was a thirty-nine-year-old Canadian and a veteran of the First World War. He was born in Iceland on November 9, 1891, but by the age of eight or nine had moved to Arborg, Manitoba, with his parents, Gunnar Gudmundsson and Veronika Eiriksdottir. He grew up on his parents' farm but left to join the military when Canada entered World War I. He served in the Canadian Expeditionary Force (CEF), but little else is known about his war experiences. The CEF was composed of several combat formations, the largest being the Canadian Corps. The corps was particularly active on the Western Front in France, in places such as Vimy Ridge and Passchendaele, and in Belgium at Ypres. Hiring records at the time of the Waldron Creek Fire indicate that Gunnarson had just recently arrived in Great Falls from Lethbridge, Alberta. Interestingly, an insurance policy found in a hotel room he had occupied before leaving for the fire bore the name Gudmundsson.

Herbert Novotny and Frank M. Williamson, best friends from Great Falls, stepped up to the challenge as well. Novotny, twenty, was born in Belt, Montana, on January 28, 1911. One of eight children, he was the son of John L. Novotny and Mattie Byers, a mixed-race couple. Cascade County birth records for the years 1893–1917 indicated that he also had a twin brother. While these records stated that his color was "W" (white), the 1930 United States federal census identified Herbert as "Negro" under the race category. Herbert was married to Agnes M. Williams, and together they had two young daughters. According to *Polk's Great Falls Directory* of 1929–30, Herbert, whose occupation was listed as "trucker," and his young family lived at 1303 Sixth Avenue South. As noted by a local historian:

> *From its founding in 1884, Great Falls practiced a "soft" form of segregation, with African American residents restricted to the lower Southside. Over the years this segregation became more rigid, with blacks excluded from union membership, jobs at the copper smelter and rail repair yards, barbershops, nightclubs, and restaurants. Blacks worked downtown in service industries or for the Great Northern Railroad, worshipped at the Union Bethel African Methodist Episcopal (A.M.E.) Church, lived in black hotels, the porters' quarters of a few family homes, and sought entertainment at "colored clubs."*[19]

Williamson, twenty-four, was born in O'Neill, Nebraska. At the time of the fire, he had lived in Great Falls for a period of about nine years, picking up various jobs to make ends meet. He was a cook at the Quick Lunch in Great Falls and also an employee at the Pastime Pool Hall there. His mother lived on Vaughn Road, just outside of town, with Frank's stepfather, a car inspector for the Great Northern Railroad by the name of Frank Blasing.

Ted Bierchen, forty-six, also stepped forward. Born in Luxembourg on October 25, 1884, he immigrated to America in 1902, when he was eighteen. As of 1910, he was staying with his brother John, working at his greenhouse business in Illinois. Later, Bierchen was reputed to be heading to California due to the scarcity of jobs in the Chicago area. His World War I draft card listed his residence as Casselton, North Dakota, and noted that his present occupation was "farming." His draft registration card was not signed until September 1918. Since the war officially ended in November 1918, chances are he was never called up to active duty, but that is mere speculation. For some reason, by 1930 Bierchen ended up in Great Falls, where he found work as a laborer on a dairy farm. It remains unknown what circumstances prompted him to seek employment as a firefighter.

The last of the five to join the group was Charles Allen, age about thirty-seven. His hiring records for the Waldron Creek Fire listed a home address in Pittsburgh, Pennsylvania. After his death, the *Great Falls Tribune* noted that friends of Allen commented that he, too, was a veteran of World War I. However, a search of U.S. draft registration cards for the war period using the name "Charles Allen" results in 1,676 matches, so without more specific search criteria, reliable information about his military experiences is nearly impossible to obtain. Friends of Allen also noted that he worked for the Sells-Floto circus for a period of time. The Sells-Floto circus was a combination of two touring acts, the Sells Brothers Circus and the Floto Dog and Pony Show. So if you have ever heard of something being referred to as a "dog and pony show," its origin can be traced here. Buffalo Bill Cody headlined the circus during the 1914 and 1915 seasons.

Again, according to the testimony from some of the fifteen individuals who survived, Gunnarson took the lead. By their account, the Canadian argued that the men were in an unsafe position and that he had much experience with fighting fires. We know that Novotny, Williamson, Bierchen and Allen decided to throw in their lots with him. Leaving the others, the five men worked their way to the bottom of the gulch, into the path of the fast-approaching main fire. What remains unknown is if the men were in full retreat or had instead decided to go on the offensive against the blaze. If they had resolved to flee, it seems odd that they did so in the most dangerous direction: toward the quickly expanding main fire.

The other fifteen men retreated to a previously burned area, away from the oncoming blaze. They climbed uphill and along the ridge before escaping down the Middle Fork of the Teton River to its confluence point with the North Fork. Whether this was a well thought out and executed plan or simple dumb luck cannot be determined. Eventually, they stumbled their way into the fire camp, which had been moved some distance up the North Fork for fear that the fire might burn over the original camp location.

On Wednesday morning, a roll call of the recruits revealed what most of the men already knew: five members of their crew were missing. District forest ranger Streeter and alternate ranger McNeal organized a search party. For reasons still unknown, the bodies of four of the men were not found until the next day, Thursday. Huddled together near a freshly dug trench and what appeared to be a constructed fireline, searchers found the charred remains of four individuals. Although they were burned beyond recognition, identification was possible by the size of each man, items found in his pockets

Forest Service men with livestock. Date unknown. *University of Montana Archives and Special Collections.*

and what remained of his clothes. In this manner, the bodies of Gunnarson, Bierchen, Allen and Williamson were positively identified.

Following the discovery of the bodies, word was sent immediately to Teton County coroner Charles G. Roberts via the Wrong Creek and Ear Mountain stations. Roberts was an experienced mortician, having practiced in various parts of the state for many years. He had been coroner since being elected to the post in 1929, after moving to Choteau and opening Roberts Funeral Home. In short order, Roberts and six others, traveling in two vehicles, drove as far as the Deep Canyon Ranch and then hiked the remainder of the way to the fire with the help of livestock.

Upon Roberts's arrival on the scene, he gave orders that the search be extended for the missing fifth man. It was not until the next day, Friday, that the body of Novotny was finally found, approximately seventy-five yards farther up the slope from the others. Roberts determined that both Gunnarson and Allen had broken their legs. He surmised that this had occurred as the men ran through the rocks and downed timber. Coroner Roberts, however, said that it was also possible that the broken bone in the ankle of one of the men might have been caused by the drawing up of the foot as a result of exposure to the heat of the fire. Roberts also theorized that the men might have become unconscious from exhaustion, heat and smoke before the flames actually reached them.

Coroner Roberts and his six men struggled to remove the bodies from the roughly 7,500-foot elevation where they had fallen. Using litters and pack horses, it was late Saturday afternoon before the group made its way out of the fire and down to the road where they had left their vehicles. Once there, they commenced the somber drive back to Choteau. Already, with the bodies of the men barely cold, the Forest Service was figuring out how to cover its ass in terms of being responsible for their deaths. Fortunately for the organization, it was fairly well practiced in such maneuvers.

3

THE AFTERMATH

The fire in Waldron Creek was far from being the only troublesome blaze that the Forest Service had to contend with during this period of time in late August 1931. In Idaho, 600 firefighters were attempting to save the towns of Grimes Pass and Pioneerville. The tiny hamlets of Quartzburg and Granite Creek had already been almost completely destroyed when fire roared through them. Kellogg and Wardner were endangered, as well, by unchecked conflagrations. The McPherson Fire in the Coeur d'Alene Forest was being fought by over 1,700 men. In Montana, thousands of firefighters were doing battle against blazes in each of the national forests there. The Deer Creek Fire in the Kootenai Forest had nearly 1,500 firefighters on the line. A blaze southwest of Whitefish nearly burned through town, and only a last-minute wind shift averted the disaster.

Tales of terror and heroism were pouring in from the firelines. As fire—or what newspapers had taken to calling the "red demon"—swept down on his small Idaho ranch, Ed Dailey buried his two small children in a potato patch, placing wet sacks over their faces to keep them from suffocating. They survived the night, although their house, outbuildings and most of their belongings did not. Wild animals perished in the onslaught of fire, even the fleetest of foot. On farms, chickens were roasted alive in their coops. Ten-year-old Marvin Jackson, barefoot and terrified, helped to guide his mother and a group of other children, including a one-day-old baby, through the flames to safety. At times, he had to swat out burning embers in the blanket that covered the small infant.

The Forest Service was waging an all-out war against fire, and in many cases it appeared to be losing. To make this combat analogy work required actual casualties, and these had begun to mount in earnest. Besides the five killed in Waldron Creek, two other men were burned to death on August 25 near the South Fork of the Payette River in Idaho. In the Kootenai Forest in Montana, a fire lookout was killed by a falling snag on August 26. Dozens more firefighters were being treated at various hospitals for ailments ranging from cuts to eye injuries to food poisoning. Mickey West, "one of the best known woodsworkers of western Montana,"[20] was crushed to death by a rolling rock on August 30. He and another man were sawing down a tree on a fire nineteen miles east of Hamilton when the accident occurred. While the other man barely escaped the mishap, West was not so fortunate.

Although the fires were wreaking havoc on the landscape and on a sizeable number of those men who were fighting them, they were having a strangely positive effect as well: providing work for the unemployed. As noted earlier, the Great Depression was in full swing, and thousands upon thousands of people in Montana, Idaho and elsewhere were out of work. The local Missoula newspaper commented:

> *The sudden outbreak of forest fires means relief for the moment from the long search for a job. Truckloads of men, their faces beaming as if they were on the way to a picnic instead of the long hours and dirty, grinding toll of the fire lines, roared into the woods from the city...Early in the summer when the first installment of the 1931 forest fires appeared the flow of jobless men started to Missoula. From as far east as Chicago they came as the word, "Forest fires out in Montana," was passed along the grapevine telegraph of the road. All sorts were represented in the hundreds congregating daily around the Forest Service employment office on Rose Avenue. Boys from drouth-burned* [sic] *and grasshopper-infested farms, spectacled clerks, artisans unable to find work at their trades—all working types flocked to the Garden city hoping to obtain a few days' or weeks' pay in the fire trenches. A tough-looking lot, they are. Not a linen collar in the crowd. Still, likely as not, an unshaven tramp picked at random may turn out to have been a year or two ago a respected citizen in his home town, attending church occasionally, keeping up his lodge dues and perhaps paying installments on a car. Some family men are among the firefighters, hoping to send a few dollars home to the wife and kids.*[21]

Unfortunately, it was not just Mother Nature and her lightning storms that were responsible for all the new fires that were being started. Men, many of them firefighters looking for more work, were setting off blazes almost as fast as their brethren could put them out. The government dubbed them "firebugs" or "incendiaries." The problem became so pronounced that on August 29, Idaho governor C. Ben Ross proclaimed martial law in four counties and ordered out Idaho National Guard troops to suppress the arsonists. The four counties were Gem, Boise, Valley and Idaho. Five rifle units, a machine gun company and two medical units were activated, over two hundred men in total. Entrance into any national forest within these four counties was prohibited except on a permit, which was issued only by the governor or other designated authority.

Governor Ross called them "drastic measures" but noted that they were necessary to "curb the work of men who maliciously light fires in order to give themselves employment in fighting them." He elaborated, "It is a vicious practice that some men have fallen into. It has hindered control of the fires. Hardly is one fire under control than another started. Somebody is doing it. We know their reasons, of course, are to keep themselves employed...We will hunt them out."

Immediately upon issuance of martial law, a "cleanup campaign" was started, with particular focus on the firefighters on the Boise Basin–Payette River front. Forty-five men were identified as "undesirable" and were discharged by the Forest Service and escorted under guard out of the restricted area. It is not known what criteria were utilized to label someone "undesirable." The law under which martial law was proclaimed provided for a minimum penalty of two years in prison for anyone resisting the authority of the military within the prescribed areas. On September 1, Governor Ross placed two more counties, Adams and Lemhi, under martial law and called out extra troops to patrol these additional areas in danger of incendiary fires. Days later, Custer County was added to the list as well.

Although the arson problem was evident in Montana, the state resisted calling out the National Guard. Instead, it relied on its law enforcement officers and justice system to quell the incendiarism by making an example of those caught. Joe Brady, a firefighter from St. Paul, was charged with throwing live embers outside the fire line at the McLeod Peak Fire at the head of the Rattlesnake area in the Lolo Forest. He appeared before United States commissioner Ward H. Jones for arraignment. Bond was set at $500, an astounding amount of money at the time, which Jones was unable pay. He was remanded to the custody of the sheriff and thrown in jail, effectively thwarting his ability to start any new fires.

The aggressive responses to the burning fires, be it the thousands of firefighters employed to fight them or the declaration of martial law to keep firebugs out of the woods, could truthfully be attributed to one primary reason: trees. The Forest Service was losing timber to the blazes, timber the agency was desperately trying to protect so it could sell or harvest those board feet later. The cold, hard truth was that, in the eyes of many different organizations, not just the Forest Service, and the individuals working for them, the timber was far more important and valuable than the people attempting to put out the fires.

On September 1, 1932, Forest Service chief Robert Stuart submitted his annual "Report of the Forester." Covering the fiscal year that ran from July 1, 1931, to June 30, 1932, it encompassed most of the fire season of 1931. The Waldron Creek Fire was included in this time period. This report went to not only Arthur M. Hyde, secretary of agriculture, but the U.S. Congress as well. The opening section of the document is titled "New Light on the Timber Situation." The first words of the report state:

> *That the timber supply of the United States is waning has long been generally believed. The Forest Service has for many years been diligently seeking better light on how much timber there is in the country, how fast it is being used up, how fast it is growing, and how use and replenishment can be brought into proper balance.*[22]

By the end of the 1931 fire season, twenty firefighters would be killed fighting Forest Service blazes. This fact would be noted in the report, but not until page fifteen, after Stuart had covered such topics as cost accounting, cooperation with states in tree planting and land acquisition. Not even a paragraph in length, the entry about the deaths simply says:

> *In 1931 twenty Forest Service employees were killed on fire-suppression work. Their names follow:*

Herbert Harney.	*Ed Murphy.*
Nolan Warner.	*Carl A. Obermeyer.*
James Taylor.	*V. Leroy West.*
Herbert Novotny.	*J. Kane.*
Frank Williamson.	*Ira Hayden.*
Ted Bierchen.	*Vernon L. Cresson.*
Hjalmer Gudmundsen.	*Frank E. Jones.*
Charles Allen.	*Jesse B. Paige.*

Raymond Helm.	*Chas. E. Bramhill.*
John Moss.	*Louis Pariseau.*

That is the entirety of it. Some of the names on the list are incomplete or misspelled, and there is no clarification of how or where they died. The report then goes on to describe, in page after page of painstaking detail, the number of fires for the fiscal year, their size and cause, the number of acres burned, the cost to fight them and the total damage due to lost timber. Twenty men were killed fighting its fires—the highest loss of life since the Big Blowup in 1910—yet all the Forest Service could devote to the dead in its annual report was about two inches of space in a thirty-eight-page document.

Harry Gisborne worked as a fire scientist with the Forest Service for a period of nearly thirty years. Dubbed by the agency as "the first true specialist in forest fire research in the Nation," Gisborne invented or modified numerous instruments to measure fire conditions and was a fierce proponent of gathering data directly from actual forest fires. He was also an advocate of publishing many of his findings, an act that was uncommon to other investigators of the time. Subsequent fire researchers have called the time period from 1922 to 1949 the "Gisborne Era" due to his wide-reaching influence. Gisborne died in November 1949 in Mann Gulch, the place where twelve smokejumpers and a recreation guard died while fighting that tragic fire in August of the same year. True to his scientific roots, he succumbed to a heart attack on the actual fire ground while checking some of his theories on fire behavior.

Gisborne compiled an annual report each year detailing that year's fire season in the Northern Rockies region. "The Character of the 1931 Fire Season in the Northern Rocky Mountain Region" was submitted by Gisborne on October 2 of that year. In the opening paragraph, he stated, "Soon after the close of each fire season every timber protective organization in the northern Rocky Mountain region is confronted with the need of rating its success in protecting its timberland resources from fire." As opposed to rating the achievements of a particular fire season based on a factor such as the safety of those fighting the fires or the number of lives lost on fires, Gisborne instead noted that fire seasons should be evaluated by how many board feet of timber were saved from being burned.

It was not just the Forest Service that could be accused of placing more value on trees than human life. Idaho governor Ross, when talking about his declaration of martial law and the fact that "firebugs" looking for employment were starting many of the blazes around the region, commented, "My sympathy goes out to any man out of work, but there is certainly no sympathy shown to those who would sacrifice the rich timber, burn up our towns, and endanger our people in order to provide themselves with wages." Intentional or not, Governor Ross's statement clearly listed the welfare of the timber ahead of that of the people and hinted at where his true priorities resided.

With regard to Waldron Creek, it appears that timber also played a direct role in the Forest Service's decision to actively battle this specific blaze. Earlier in the year of 1931, the agency had sold the timber in the Waldron Creek drainage to Nels Stengrumson, a local lumberman. Stengrumson had even gone so far as to build part of a road leading to the site and was planning on moving a sawmill there. The Forest Service was financially committed to protecting these trees since they had now been sold off to a bidder. With the timber under contract, it is fair to assume that the agency was more willing to aggressively fight the fire, even if it meant putting firefighters at an increased risk. Shortly after the fire was controlled and mopped up, district ranger Streeter estimated that no more than 30 percent of the timber was damaged. Although many of the trees were dead, it was Streeter's estimation that most of the timber could still be harvested.

After erupting and claiming the lives of our five men, the Waldron Creek Fire languished. As if exhausted after its fateful run, it lay down and refused to move much more, remaining at about eight hundred acres in total size. Twenty-five men had been removed from the fire after the blowup on Tuesday, August 25. By Wednesday, 50 more had been sent in to begin mopping up and looking for the bodies of the missing 5 firefighters. As noted previously, four of the bodies would not be found until Thursday the twenty-seventh, and the fifth would be found on Friday the twenty-eighth. At this point, about 125 total personnel were assigned to the blaze. They would continue mopping up hot spots for about the next week until being released. The total cost for suppressing the fire ended up being $12,424, a considerable amount of money at the time and nearly $3,000 more than the Teton District's entire yearly budget.

The remains of the five firefighters were brought directly to Roberts Funeral Home in Choteau late on Saturday, August 29. A hearse from the T.F. O'Connor Funeral Home in Great Falls was dispatched to retrieve the body of Novotny, while Williamson's remains were brought back to town by the Merrill Mortuary. The two were soon laid to rest in Highland Cemetery on the outskirts of Great Falls. Within a couple days, the body of Ted Bierchen was shipped to Chicago for funeral services and burial.

Ted's brother Jake was in charge of those arrangements. The bodies of Allen and Gunnarson were held in Choteau as the process of contacting relatives continued. Eventually, they were buried in unmarked graves in the Choteau Cemetery.

SPIN IS A FASCINATING WORD. As a verb, it can mean many things: to whirl around quickly, to draw out and convert into threads or even a process of fishing. As a noun, it can refer to the revolving motion imparted to a ball in a game such as baseball, the uncontrolled fast descent of an aircraft resulting from a stall or even the intrinsic angular momentum of a subatomic particle. And when used in a more informal manner, it can mean a particular bias, interpretation or point of view intended to create a favorable impression when presented to the public. It is on this latter definition that we shall now focus our attention.

Realizing that they had five deceased men on their hands and that they might have some degree of accountability for those deaths, Forest Service officials and others did their best to put some spin on what happened. Coroner Roberts was apparently more than willing to help. An inquest was held in Waldron Creek on August 28, shortly after finding the fifth and final body—that of Herbert Novotny. In the Teton County Coroner's Register, the official documentation for all deaths in the county, Roberts compiled five separate entries, one for each man.

Under the column "Names of Jurors," he wrote, "No Jury called." For "Names of Witnesses" to the inquest, he listed Freeman Page, Franklin Fellers, Clyde Darby, Bud McNeal and Dick Hendricks. Page was a smokechaser in the Teton District and the first to actually spot the Waldron Creek Fire from the lookout tower on Mount Wright on August 23. Fellers was the local farmer turned guide who had led the twenty men to the spot fire and then abandoned them while he went to tie in with McNeal, the alternate district ranger, who was on the main fire. Darby, who would eventually become chief of police in Choteau, and Hendricks were also members of the twenty-person crew to which the victims belonged. Let us just say it was a convenient sampling of witnesses, all agency men or at least close enough to the agency that they were willing to toe the company line if necessary. In this case, it was most certainly necessary. Five men were dead.

Lastly, in the Coroner's Register section titled "Findings of Jury as to Cause of Death," Roberts wrote in each man's entry: "Came to his death by being traped [*sic*] in forest fire. No one to blame but himself." Blaming the

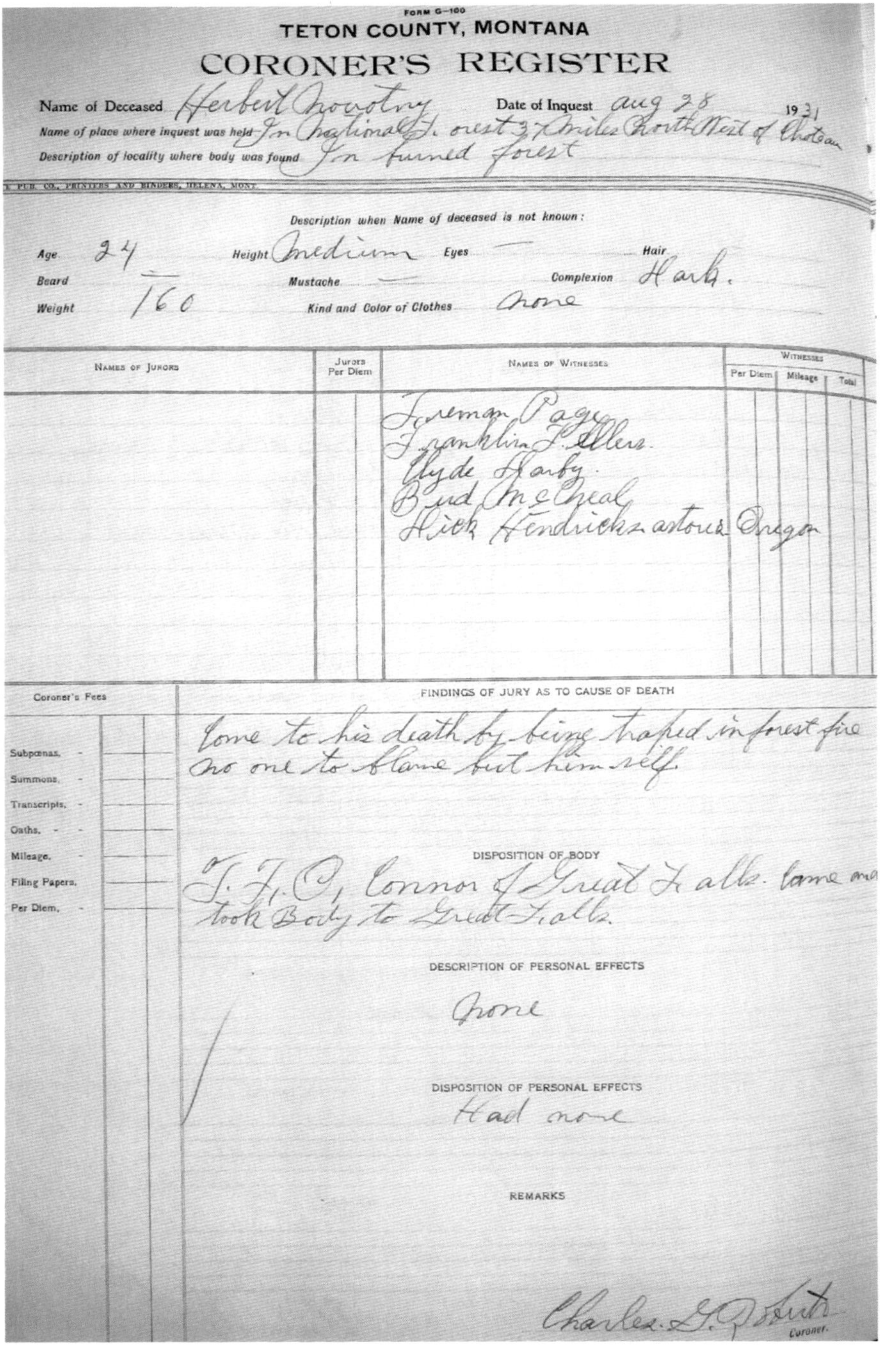

FORM G-100

TETON COUNTY, MONTANA

CORONER'S REGISTER

Name of Deceased Herbert Novotny Date of Inquest Aug 28 1931

Name of place where inquest was held In National Forest 37 miles North West of Choteau

Description of locality where body was found In burned forest

Description when Name of deceased is not known:

Age 24 Height Medium Eyes —— Hair

Beard —— Mustache —— Complexion Dark.

Weight 160 Kind and Color of Clothes None

Names of Jurors	Jurors Per Diem	Names of Witnesses	Witnesses Per Diem	Mileage	Total
		Foreman Page			
		Franklin F. Ellers.			
		Clyde Darby.			
		Bud McNeal			
		Hick Hendricks astoria Oregon			

Coroner's Fees: Subpœnas, Summons, Transcripts, Oaths, Mileage, Filing Papers, Per Diem

FINDINGS OF JURY AS TO CAUSE OF DEATH

Come to his death by being traped in forest fire no one to blame but him self.

DISPOSITION OF BODY

T. F. O, Connor of Great Falls. Come and took Body to Great Falls.

DESCRIPTION OF PERSONAL EFFECTS

None

DISPOSITION OF PERSONAL EFFECTS

Had none

REMARKS

Charles G. Roberts
Coroner.

Coroner's Register, Choteau, Montana. *Author's collection.*

victim has long been a popular strategy, from the days of sinners in the Old Testament to today's unfortunate targets of racism and rape. Sadly, part of the reason victim blaming has such a lengthy history is due to the fact that it has proven over and over again to be an effective tactic. It seemed to work here as well, since no one questioned his assignment of guilt to the deceased. For his efforts, Coroner Roberts submitted his bill: $7.40 for mileage, $15.00 for three days' worth of work. For the Forest Service, it was $22.40 well spent.

To build the paper trail of defense a little longer, Teton District ranger Walt Streeter traveled to Great Falls on September 3, nine days after the men died in the fire. In his pocket he carried with him a statement already typed out. All it needed were some signatures. It read:

> *On Tuesday, August 25, 1931, we, Pat Menehan, Tom Sullivan and Dick Hendricks, firefighters working on the Waldron Creek Fire on the Lewis & Clark National Forest, were members of a crew, five of whom met with accidental death on the fire upon that date.*
>
> *Ranger W.N. Streeter, in charge of the fire, sent us out to work in charge of a guide named Franklin Sellers, who upon instruction of Ranger Streeter took us to the lower edge of a twenty acre burn about a quarter of a mile east of the fire, and left us there while he went to give the foreman named C.A. McNeal, a note instructing him where to put his crew to work.*
>
> *While the guide was gone, the fire began to crown and advance toward us. All of the crew except the five men who were burned ran into the burned area and thence up to the top of the ridge. Instead of coming with us the five went down the slope toward the fire, and were overtaken by it, causing their deaths.*
>
> *It is our opinion that all possible precautions were taken by those in charge of the fire to provide for our safety.*
>
> *In witness of the foregoing statement, we, each of us, hereunto subscribe our names, on the 3rd day of September, 1931.*

No one seemed to question why the men were left on the spot fire without any leadership or supervision. Nor did they ask how it was possible that, if all precautions had been taken by those in charge, five men were still dead. For some reason, perhaps due to haste or ignorance, the name of the guide was listed as "Sellers" when in fact his actual name was "Fellers." The three men affixed their signatures to the statement, as did Streeter and a witness by the name of D.M. Preston just for good measure. With this documentation, the "investigation" into the five deaths on the Waldron Creek Fire was over.

However, not everyone was to remain silent about the loss of life in Waldron Creek. E.L. Jourdonnais, owner and publisher of the *Choteau Acantha*, wrote an editorial in the September 3 edition of his own paper to decry the fatalities. He stated:

> *Men who are killed in action while serving with the United States army, navy or marines are reported as having died in the line of duty. There is a particular sentiment attached to those few words which is most appreciated by those who have been enlisted men and perhaps have seen some of their comrades stricken down.*
>
> *It seems fitting that this term might be borrowed and be applied to the five men who last week were burned to death while fighting the forest fire on Waldron creek west of Choteau. While theirs was another duty, they truly risked their lives and lost. Such a tragedy brings to our minds the hazards which men dare when they go into the great forest to battle with a blazing inferno. When they lose, it seems fitting that some form of public recognition should be paid them for the supreme sacrifice they have made.*
>
> *Fire fighting forces as a rule are hastily recruited. The exigencies of the occasion necessarily make it so. And for that reason it is hardly to be doubted but that men are enrolled in the ranks of fire fighters who are lacking in experience and, perhaps in frequent cases, in physique. With the magnitude that forest fire fighting had grown to in recent years, there seems to be justifiable reasons why the state and national governments may well give more systematic attention to the mobilizing and training of fire fighting crews. It even is not beyond the bounds of reason to suggest that the national and state governments maintain regular standing armies of fire fighters during the height of the fire season, and that these men should be trained and submitted to physical tests as in the case with armed forces. It, of course, goes without saying that adequate wages should be paid these men.*
>
> *Two desirable objectives might be accomplished by this course. One is that* [there] *would be less likelihood of the repetition of such tragedies as occurred in the forest west of Choteau, though it is not to be argued that fatalities would be eliminated altogether. The other is that efficiency would be improved. There seems to be a widespread belief at the present time that incendiarism has been a large factor in the forest fire situation this year. Men are accused of setting fires to provide themselves with jobs fighting them. This situation became so acute in Idaho that Governor Ross*

of that state placed three counties under martial law. He declared that the situation called for drastic measures. The action of the Idaho governor gives a suggestion to an idea which it seems may be feasible for other states as well as the national government to adopt. If 100,000 men in our national army, augmented by a large number from the national guards of the state, could be put on fire patrol duty in the forests during the summer it does not seem unreasonable to believe that a mighty barrier could be raised against the danger of forest fires. It seems that the situation has reached a status where, in the words of Governor Ross, drastic action is called for.

These thoughts come anent to the tragedy on Waldron Creek. We started out to eulogize those men who died there, and so we want to conclude. They are real heroes and their lifeless forms are entitled to the caresses of Old Glory. Every consolement should be offered to the bereaved relatives and friends for there may be more personal tragedies about which the world will never know. Nay more, the national government owes it to the survivors of these men to be solicitous of the welfare of those who may be dependent upon them or might be in the near future.

Unfortunately, the Forest Service and the United States government felt no need or desire to be "solicitous of the welfare" of the victims' dependents. They would not cover the cost of headstones for the graves, let alone pay survivor benefits. Nor would they offer any type of "consolement" to the bereaved friends and relatives. These were not actions the government had taken in the past when firefighters died under their watch, and it had no desire to set that precedent now. Interestingly, though, it would not be long before one of Jourdonnais' other suggestions came to fruition.

At this time, as noted by the *Acantha* publisher, there were no "standing armies of firefighters" trained and equipped by the federal government.[23] Each district in the Forest Service that contended with fires hired a small number of smokechasers and fireguards each year. But especially during busy fire seasons, this number could never begin to meet the needs in the field. Thus, for decades the government had come to rely on what has been referred to as the "fire militia" for meeting its firefighter personnel needs. Since its inception, when things got busy, the Forest Service, like it had done when looking for people to fight the Waldron Creek Fire, simply hired folks off the streets who were looking for work. For decades, this had proven to be a quick and reliable way to hire firefighters. The drawbacks of such a system were that the men hired had widely varying degrees of fire experience. Most of them probably had no formalized training in how to fight fires safely, they did not have to pass any physical fitness requirements to be hired and there was little, if anything, in terms of a defined chain of command among those employed. All of these factors played some role in the deaths of the five in Waldron Creek.

Slowly, though, after the firefighter fatalities of 1931, things began to change. Launched in 1933, and running until 1942, the Civilian Conservation Corps (CCC) was one of the first programs rolled out as part of Franklin D. Roosevelt's New Deal. In the CCC, unmarried men between

CCC boys outside Barracks 4 near Ballantine, Montana. *Photo courtesy of CCC Legacy Project.*

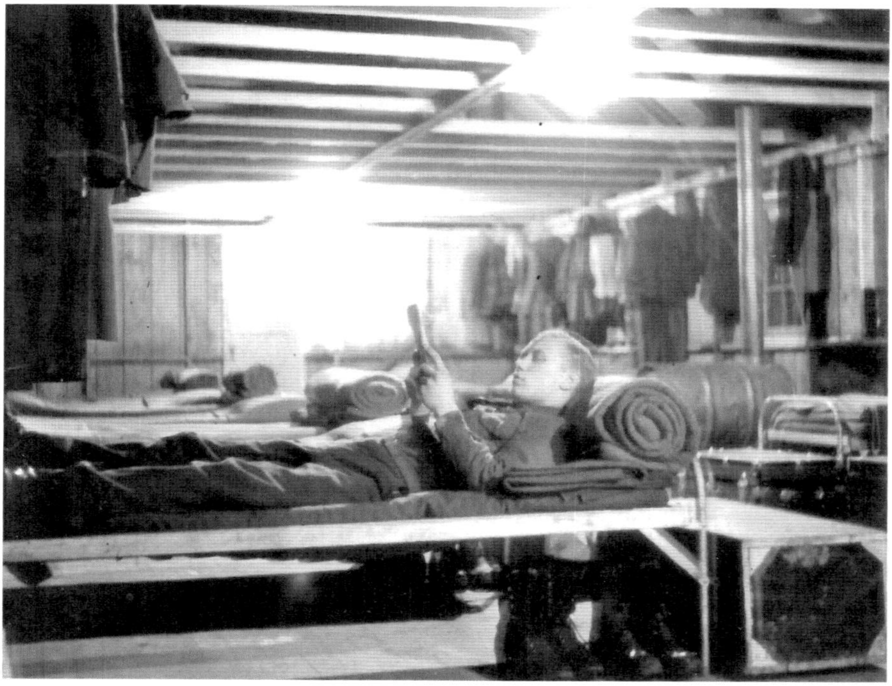

A CCC boy getting some quiet time inside his barracks. *Photo courtesy of CCC Legacy Project.*

CCC firefighters recuperating after battling a fire in the Bitterroot National Forest. *Photo courtesy of CCC Legacy Project.*

the ages of eighteen and twenty-three were put to work on projects in rural areas that were typically owned by federal, state and local governments. Due to its popularity, the program was eventually expanded to include men ages seventeen to twenty-eight. The guys usually served a term of six months but could extend their service up to four terms. They earned thirty dollars a month, with the stipulation that twenty-five dollars of it be sent home to their families. CCC enrollees built hiking trails, roads and buildings; they constructed bridges and planted trees; and, most relevant to this discussion, they worked on fire crews. Finally, in 1933, for the first time, the Forest Service had at its disposal a large workforce or "standing army" of firefighters that it could rely on when needed for fire suppression.

However, just because these CCC crews were trained as firefighters and functioned in this capacity during the summer months, it did not make them immune to their own tragedy on the fireline. On August 21, 1937, ten CCC firefighters, along with five Forest Service employees, were killed in the Blackwater Fire on the Shoshone National Forest near Cody, Wyoming. Thirty-eight other firefighters were injured. The CCC enrollees, some of them still too young to shave, ranged between the ages of seventeen and twenty. An undiscovered spot fire blew up when a dry cold front and its accompanying high winds passed over the area. Fire raced uphill and overtook the men, leaving in its wake over fifty dead and injured. It stands as the largest loss of life in a single fire between the Big Blowup of 1910 and the nineteen who were killed on the Yarnell Hill Fire in 2013.

In 1939, on an experimental basis, the first organized Forest Service fire suppression force was created.[24] Dubbed the "40-man" crew, it was located on the Siskiyou National Forest in southwestern Oregon[25] and proved to be highly effective in fighting fires in this part of the country. The program quickly expanded, and by 1947, the Forest Service had begun organizing "Hotshot" crews for fire suppression work specifically in California. The "Hotshot" moniker comes from the fact that these crews were typically assigned to the hottest, most dangerous parts of the fire. By 1961, the first Interregional (IR) fire suppression crews were created. These were twenty-five-person teams complete with a foreman, an assistant foreman, three squad bosses and roughly twenty crew members. These IR crews were highly mobile and able to be utilized wherever needed around the West. Eventually, these IR crews assumed the name Hotshots and reduced their size by five members to be more efficient. Today, there are about 110 of these twenty-person Hotshot crews around the country.

Also in 1939, the Forest Service's Aerial Experimental Project was moved from California to Washington State, and its focus shifted from experimentation in the use of water and chemical bombs for fire suppression to experimenting with parachute jumping. The men involved in this program were called smokejumpers. The smokejumper program proved to be highly feasible, and by July 1940, Rufus Robinson and Earl Cooley had made the first operational fire jump in Martin Creek in the

The early days of smokejumping. Sleeman Creek, near Lolo, Montana. Date unknown. *University of Montana Archives and Special Collections.*

A smokejumper with parachute, circa 1944. *University of Montana Archives and Special Collections.*

Nez Perce Forest in Idaho. The smokejumper program likewise expanded rapidly due to its demonstrated effectiveness and ability to keep many fires small and thus cheaper to extinguish. Currently, there are nine smokejumper bases located throughout the western United States and over four hundred operational smokejumpers.

So within two years after the deadly fire season of 1931, the Forest Service had at its disposal thousands of organized, equipped and basically trained firefighters in the CCC. By the end of the decade, it would also have available a small number of 40-man crews, as well as a modest quantity of highly fit, well-trained smokejumpers in both Washington State and Montana. While it would continue to rely on this "fire militia" strategy for years to come (and still does today, to an extent), the Forest Service obviously decided that it needed to hire, train and equip a "standing army" of firefighters, just as Jourdonnais suggested in his paper's opinion editorial section. The deaths of twenty men over the course of the 1931 fire season, including the five in Waldron Creek, must have played some role in influencing this decision. Despite being forgotten, the five men who died in Waldron contributed to the legacy of the modern-day wildland firefighter.

By September 9, 1931, a minimum one-half inch of precipitation, in the form of rain and snow, had fallen over the entirety of the Northern Rockies region. Martial law in the Idaho forests was rescinded. The troops stationed in the woods were withdrawn since forest officials felt comfortable that the fire hazard was greatly reduced and that regular forces could take up the patrol duties handled by the troops. R.H. Rutledge, intermountain regional forester, said that higher humidity and cold nights were preventing fires from "getting headway." On September 16, Ranger Streeter laid off the small number of smokechasers, fireguards and other fire-related personnel who had been hired in the Teton District. About fifteen others who were put to work fighting fires were returned to their regular duties in the forest, which included maintenance and improvement work. The deadly fire season of 1931 had finally come to an end in the Northern Rockies.

Interlude

The Guts of the Monster

He's living there, in the disembodied bowels of the now dead beast. In this case, that ravenous monster was the former Polley's Mill in the center of Missoula, which chewed up logs and spit out an assortment of timber-related products. The innards are parts of what kept the place alive: a propeller shaft that served as the heart of the circulatory system for heating; twin turbochargers—lungs if you will—for processing expired gases; a small teepee burner to help excrete the waste; and a network of ductwork entrails that joined most of the whole mess together. At some point in the not-so-distant past, these insides from the mill had been discarded here, in an as-yet-to-be-salvaged portion of the abandoned site. While certainly parts of the Old Sawmill District have been reclaimed, there are other sections, like here, that remain untouched, a work in progress—ruggedly handsome from a distance but rough around the edges upon closer inspection, much like the men who used to work here.

It is hard to believe that someone would decide to take up residence in these castoffs, exposed to the elements like they are. Near as I can tell, he has been living here for a long while. I first saw his bike and a mattress there in October 2014 during one of my many trips to Silver Park. I checked again in early December, when several inches of fresh snow blanketed the ground. Sure enough, leaning up against the rusted-out guts of the dead mill was the same bike, its rear reflector catching the last rays of a fading sun.

In January of the next year, I saw tracks in the snow heading down to the metallic entrails. I followed them, exhilarated and scared shitless at the same time. As I got closer, I saw the bike and some grocery store plastic bags filled with what appeared to be clothes and other unidentifiable personal belongings. I thought about a closer inspection, but suddenly my risk radar went off, prompting me to decide that the best course of action at the moment was a hasty retreat. I withdrew to the presumed safety of the pavilion, parked myself on a wooden picnic table and observed the steel viscera from afar. Soon enough, I saw him emerge, dressed in black pants and an equally murky sweatshirt, the

The discarded ruins of Polley's Mill. Missoula, Montana, October 2014. *Author's collection.*

hood of which was drawn low ever his eyes. He began walking to the north, on a collision course with the wooden structure where I sat. I contemplated a withdrawal to the south, but fear and curiosity froze me to my seat. He walked by at close range, and I forced myself to look at his face. Although most of his features were obscured, I did see his eyes, inky black, through the veil of the hoodie. We exchanged greetings. He seemed cool, engaging yet distant at the same time. He was there one moment passing by me, but before I knew it he was gone, vanished like the mirage of water on a distant highway.

About a month later, I was back in Silver Park for another of my visits. Unseasonably warm weather had devoured the snowpack, leaving the ground muddy and soft. I picked my way out to the bowels, unsure what I would find. Drawing nearer, I saw them, a cluster of four or five bicycles leaning against the side of the monster's guts. The bike closest to the contraption was the same one I had seen on my previous excursions. His bike. Either the enigmatic dark stranger was having a "house" party or the occupancy number in the structure had increased significantly. Figuring myself to be outnumbered, I went no closer.

The discarded ruins of Polley's Mill. Missoula, Montana, January 2015. *Author's collection.*

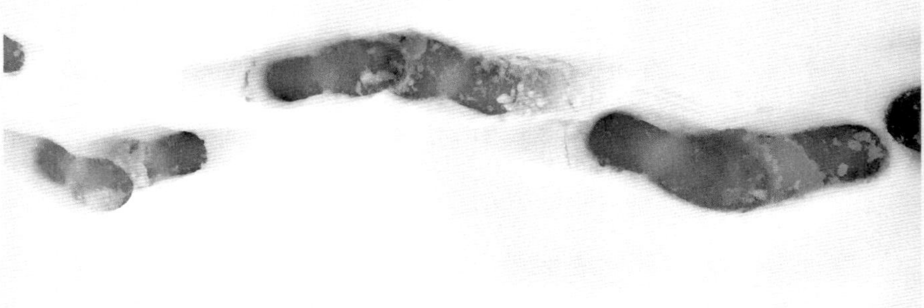

Footsteps in the snow. Polley's Mill, Missoula, Montana, January 2015. *Author's collection.*

The mill's days as a residence for whoever might be living there appear to be numbered. In June 2014, the City of Missoula put out a call for a public art project in Silver Park. With a budget of $75,000, the Missoula Public Art Committee began soliciting proposals from artists, with the intent of beautifying the fourteen-acre park space with one or more pieces of sculpture. In this proposal call, artists were informed of the abandoned relics from the old mill and encouraged to use them in their designs.

My favorite piece of the discarded ruins is a contrivance that appears to be some sort of venting shaft. In its present configuration, though, thrusting upward like it does, the resemblance to a giant phallus is difficult to deny, at least for me. It is as if parts of the old mill are saying to any who will listen, "Fuck you. We're still here. We're still vital after all these years." I would love to see how an artist might incorporate the guts, particularly this penis-like vent, into his or her design of a piece of public art. Unfortunately, that might never happen.

In the middle of February 2015, I called the coordinator of the project to check on its status. The commission for the endeavor was supposed to have been awarded on December 5 of the previous year. He informed me that an artist had indeed been selected, but that person's proposal did not include utilizing the castoff machinery. A second call for additional art projects for the park was in the works, and it was the coordinator's hope that some of these proposals would have a use for the remnants. In the meantime, he told me, the plan was to remove them altogether and keep them at the city's maintenance facility pending future developments. Whatever their fate–in a sculpture or on a junk heap or melted down for scrap–the guts of the monster were slated for removal from the park.

Therefore, I was not too surprised when, on one of my subsequent visits to the park soon after talking with the project coordinator, I saw excavating equipment working the unimproved lot where the guts lazed. For now, the machines were giving them a wide berth. Nevertheless, I could tell such a strategy would work for only so long before the insides of the old mill would have to be contended with directly.

I became obsessed with Silver Park and the dark stranger, and I followed up these thoughts with compulsive trips down to the old mill to check on him and the progress of the earthmovers. Each time I went, there he was, somewhere in the park, silently watching the heavy equipment invade closer and closer to his home. I knew that at some point, probably very soon, I would enter the park to find the guts removed, eviscerated from the vacant lot where they lay. I worried about what would happen to the man when that day came. To glimpse him on each of my trips brought a certain amount of fear, but the thought of not seeing him was even more disconcerting. If and when that time came, I knew I would struggle with understanding where he had gone and what it meant.

Due to a work commitment, I had to be out of town for nearly a week, getting back late on a Thursday evening near the end of February 2015. That Friday, after a couple hours in the office catching up on e-mails and messages, I headed down to Silver Park. Since the heavy equipment had been quite busy the last time I was here, I figured the innards might be gone. But as I approached the park on the trail from the southeast corner of the property, there they were. The big machinery was still working, mostly pushing dirt around. The guts were still lying there, in the same spot they had been when I saw them last. They were still holding on, as was the dark stranger. His bike remained, parked in full view out front, propped against one of the steel vents.

I saw him as I rode in, leaning against one of the support beams for the pavilion. He was watching, staring at the excavators and bulldozers as they crept closer and closer to the bowels. I knew it was him right away, even though I don't consciously remember processing the information to arrive at that conclusion. Somehow, I just knew. He wore a thick black coat—the cloak of a workingman, enduring in design—with the hood slumped over most of his face. I rode past. Again, I forced myself to look at his face. He met my gaze and stared right back, unblinking. Although the piercing black eyes were the same, I was surprised to see he had a dense mustache, a feature I had not seen in my previous sightings of him. I also noticed a clump of freshly dried blood on his nose, and he was missing part of a front tooth, remnants of what appeared to be a fairly recent fistfight, opponent unknown. We nodded at each other, and I continued along the dirt path.

I circled back around behind him on the park trail. Quietly, I rode down to the bathrooms in the center of the complex. When the callous winds blew from the east-northeast, like they were doing today, it was to here that I often retreated. I could sit, looking predominantly westward, and catch the afternoon rays from the sun, the building blocking the harsh breeze. There was no smell since the doors were locked during these winter months and no one was shitting in the toilets. Up against the wall, wedged between the sewage lids for the septic system, I could just watch. For what, I was never certain. I simply hoped that when I saw it, I would know. It was from this spot, months earlier, that I had first seen the dark stranger.

My plan on this day was to snap a picture of him as he watched the approaching machinery. I should have known better. Leaving the bike hidden, I took out my camera phone and peered around the corner of the building. Of course, he was looking right at me when my eyes focused down range. Holding my stare, he lit a cigarette from a wooden match, inhaled and then, after a lengthy pause, slowly released the smoke downwind. Breaking eye contact, I moved to a nearby park bench, wanting to capture in my notes what had just transpired while it was still fresh. When I finally looked up some ten minutes later, he was gone.

Curious for a better view of the guts, I locked my bike to a nearby rack and walked down for closer inspection. Discarded clothes and garbage were strewn about. The place was more disheveled than I had ever seen it. The encroaching equipment seemed to be spurring a sense of chaos, foretelling a dramatic climax not too far in the future. For now, the innards—and the stranger who lived in them—were still hanging on, still here. But the countdown clock keeping time for their removal was loudly ticking. It was going to stop—that much was certain. The only mystery was when.

Thursday, March 12, 2015

I was surprised at the suddenness with which it transpired. Here one day, gone the next. All along, I thought I had been preparing myself for this very moment: the time when I would enter Silver Park and find that the innards had finally been carried away. Today was that day. I came into the park from the west, an unusual direction of entry for my visits here. The heavy equipment was working, lots of it—the typical array of graders, dozers, front-end loaders and dump trucks. In spite of all the traffic moving about the reclamation site, I could immediately tell that something was amiss. The guts of the monster were gone.

Flustered, I ran out onto the mill site, no doubt a violation of some code or policy. Weaving my way behind the moving grader, I headed for a dump truck that was parked in preparation of the act for which it is named. The driver tilted the bed, let loose his volley of backfill and then slowly pulled away. He lowered the dump and stuck his head out the window. Not thinking clearly, I asked him where the old guts had gone. Somehow, the driver knew exactly what I was referring to. "Over to that chunk of city ground, east of the cemetery!" he hollered to me. I nodded my head in affirmation since I knew the area well. I had ridden my bike there numerous times to check out the city's official burial ground and the surrounding properties. It was exactly where the art project coordinator had previously told me they would be

The venting shaft from Polley's Mill. Missoula, Montana, October 2014. *Author's collection.*

heading. Moving that way, I navigated through the streets of the north side before ending up on Cemetery Road. I worked my way down an arterial and then spotted them out of the corner of my eye.

They looked different. I walked across a wide expanse of the grassy park, but my path to the entrails was interrupted by a chain link fence, which separated me from the lot where they now sat. Indeed, the guts did look altered. They had been taken apart, divorced into their major sections and then lined up, side by side. They looked almost majestic, in a rusted-out and flaking sort of way. The erect penis vent was still there, arching skyward, continuing its display of masculinity. Now, though, instead of the backdrop of the Silver Park pavilion, the surroundings were dominated by a large dirt mound that blocked the view of the interstate a short distance behind it.

"Can I help you with something?" a voice inquired from directly behind me. Defying anatomical constraints, my heart somehow relocated to the upper reaches of my windpipe. Mesmerized by the guts,

I had not heard the man approach. The voice belonged to Ron Regan, director of the city cemetery across the way.

"They just came in earlier today," he remarked. "What's your interest in them?"

I swallowed hard, hoping it might dislodge my ticker from my trachea. "I've been watching them for a while, down at Silver Park," was all I could muster for a reply. He seemed satisfied with the response since he didn't ask me to elaborate.

"I can take you in there if you want to look at them?" he commented. Still not trusting my voice, I nodded in approval.

Piece by piece, I examined the guts of the monster. When I got to the section of venting where the dark stranger had been living, all sense of the aforementioned majesty was quickly cast aside. It looked as if a giant alcoholic rat had been living there for some time. The mattress—dare I call it that—was actually a large, soiled piece of thick Styrofoam. Shredded bits of paper, plastic bags and other garbage were scattered about the metallic pipe. Large, empty aluminum cans of malt liquor, of varying brands, littered the venting as far as I could see. Squalor. Complete and utter destitution. Life in the guts of the monster. I could only wonder where the dark stranger might now be.

Garbage in the venting shaft, Polley's Mill. Missoula, Montana, March 2015. *Author's collection.*

After the removal of the guts of the monster, Polley's Mill, March 2015. *Author's collection.*

Even though the guts of Polley's Mill were now removed, my fixation with Silver Park endured. For days after their amputation from the park, I visited the site. A pile of fill dirt now occupied the space that the innards used to claim. While the excavating work had made the place much more visually appealing, it did not look quite right with them gone. The dark stranger was noticeably absent as well. Once, from across the park, I thought I saw him, standing west of the pavilion, staring at the now vacant field. Quickly, I snapped a picture. By the time I was able to ride over for a closer look, the man was gone, as if the ground of the abandoned mill had opened up and swallowed him whole.

Upon leaving the guts when I had last seen them across the street from the city cemetery, Ron had commented, "I sure hope whoever was living here doesn't find out where they are now. He'll be back living in there again." Silently, I hoped he was wrong. My unspoken wish was that the dark stranger would learn of their new location and move back into his old haunt. Then I would at least know where he was, knowledge that I found strangely comforting. Like all of us, Death needs a home as well, wherever that may be.

4

NOW AND THEN

On July 10, 2001, four Forest Service firefighters died on the Thirtymile Fire, about thirty miles north of Winthrop, Washington. Those killed were part of the Northwest Regulars (NWR) #6, a twenty-one-person hand crew from the Okanogan-Wenatchee National Forest in Washington State.[26] The Regulars were a collection of firefighters selected from the Lake Leavenworth and Naches Districts of the Okanogan-Wenatchee. On this assignment, the Regulars were being led by crew boss Ellreese Daniels, a twenty-plus-year Forest Service employee from the Lake Leavenworth District. Daniels possessed a fair degree of previous fire experience, having worked on two different Hotshot crews in the past. At his side was crew boss trainee Pete Kampen, also from Lake Leavenworth, a veteran of six previous fire seasons. Since this was a training assignment for Kampen, he would primarily run the crew but would do so with the supervision of Daniels. Under them were three squad bosses, each of whom was in charge of five to six firefighters. All told, there were eight rookies (first-year firefighters) on the crew, a relatively high number but one still within regulations.

The crew arrived on the Thirtymile Fire at about 9:00 a.m. to replace the Entiat Hotshots, who had been working the fire the whole previous evening. NWR #6 battled the stubborn blaze the rest of the morning and into the afternoon. At approximately 4:00 p.m., two engine crews that were working nearby radioed for assistance. They needed help with numerous spot fires. One squad of the Regulars (seven individuals in this case) responded. The firefighters encountered difficulty containing these spots, so collectively they disengaged and reconvened at a lunch spot a short distance away. While here at the lunch spot, fire activity on the main blaze increased significantly, compromising their escape route. The squad of Regulars, along with the two engine crews and the Entiat Hotshots (who had rejoined the fight after being off the clock for a few hours), beat a hasty retreat to the southwest and out of the fire zone. They radioed the remaining fourteen crew members, who were working to the northeast and up the river, and apprised them of the worsening situation. Crew boss Daniels, who was also serving as the incident commander (IC) of the fire, was in this group of fourteen. After assessing the conditions, he determined it was too risky to drive southwest like the others had done, so the crew headed up canyon, presumably away from the growing blaze.

Since the road dead-ended a few miles farther upriver, the crew was effectively cut off—by flames to the southwest and the end of the road two miles to the north. IC Daniels, with the help of an air attack plane overhead, began to assess different areas along the way as potential safety zones or spots where they could deploy their protective fire shelters if necessary. He selected a site characterized by an extensive rock scree field above and to the west of the road. The Chewuch River and a fairly sizeable sand bar ran just to the east. Daniels and the remaining thirteen members of NWR #6 unloaded from their van, assembled on and above the road and nervously watched the now approaching fire.

While they waited, two civilians, a man and a woman, arrived in their truck. Earlier that day, they had driven up canyon to a campground near where the road terminated. When they saw the smoke later that afternoon, the two decided it would be best to leave the area. The firefighters told the couple that their escape out of the canyon was cut off for now and that they would have to wait things out. The pair agreed, and they, too, began watching the advancing flames.

Based on the accounts of those who survived, the crew was not prepared for the suddenness with which the fire arrived. Burning embers rained down on them, followed by waves of heat, smoke, wind and fire. Eight members of the crew deployed their fire shelters directly on the road. The man and woman crowded into a shelter with one of the firefighters, a feat never before attempted in the history of shelter deployments. One of the squad bosses of the crew was high above the road on the rock scree field, observing the fire, when the first wave of superheated air arrived. He ran in the direction of the others on the road but could not get there before the heat slammed into the crew. Retreating, he headed back up the slope to the presumed safety of the scree. Four other crew members and one of the other squad bosses also retreated upslope into the rocks and joined the squad boss, who was already there. They deployed their shelters and waited for the firestorm to hit. It was not a long delay.

At some point during the event, two of the six retreated from the rocks and headed toward the road. One jumped into the river; the other sought shelter behind a large boulder before fleeing to the interior of the van. Although burned, each lived. The eight who had deployed their fire shelters on the road, plus the two civilians, also survived with minimal injuries. The remaining four in the rock scree field (Tom Craven, Karen Fitzpatrick, Jessica Johnson and Devin Weaver) were not so fortunate. They died as a result of asphyxia due to the inhalation of superheated products of combustion.

The deaths led to an extensive serious accident investigation by the Forest Service, the results of which were released in a report on September 26, 2001. The analysis concluded that the deaths were entirely avoidable and that they should trigger changes in the ways fires are fought throughout the country. "The Thirtymile Fire tragedy could have been prevented. This tragic accident was the result of many factors—factors that are agonizingly familiar," said Jim Furnish, the deputy U.S. Forest Service chief who led the investigation team. The one-hundred-plus-page report cited fourteen significant causal factors and five influencing factors that led to the deaths:

- *A failure by those in charge on the fire to gauge its potential danger and change tactics accordingly*
- *A failure to manage fatigue in those fighting the fire, to the point that vigilance and decision-making were degraded significantly due to lack of sleep*
- *A failure to maintain clear command and control of personnel by those in charge of the fire*
- *A lack of consideration given to identifying escape routes and safety zones*
- *A failure to identify an appropriate fire shelter deployment site*

The Occupational Safety and Health Administration (OSHA) also investigated the Thirtymile Fire. OSHA released its report in February 2002, and it identified several violations that occurred during the fire, including two "willful" violations, which are the most serious handed out by the agency short of shutting down a practice or operation. According to OSHA, "willful" violations show "intentional disregard of, or plain indifference to" federal workplace safety regulations. Also noted in the report were three "serious" violations, which occur when there exists a "substantial probability that death or serious physical harm could result, and the employer knew or should have known of the hazard."

The violations identified by the OSHA report included a failure to follow work-rest cycles previously developed by the Forest Service. One leader on the fire had been awake for nearly fifty hours when the accident happened. Also identified as a violation was the fact that no clear commander had been assigned to the fire. And, the report noted, not only did firefighters on the ground fail to heed signs that trouble was coming, but also fire managers offsite did not check in with the firefighters often enough to effectively monitor their progress.

Several other significant developments spun out of the Thirtymile Fire. In July 2002, Public Law 107-203 was passed by the 107th Congress. Sponsored by Senator Maria Cantwell and U.S. representative Doc Hastings, both of Washington State, the law states:

> *In the case of each fatality of an officer or employee of the Forest Service that occurs due to wildfire entrapment or burnover, the Inspector General of the Department of Agriculture shall conduct an investigation of the fatality. The investigation shall not rely on, and shall be completely independent of, any investigation of the fatality that is conducted by the Forest Service.*

Under this law, the goal of an investigation by the Forest Service's inspector general would be to determine if any crimes were committed. If it was established that laws had indeed been broken, then a firefighter could be criminally charged and possibly sent to prison. In 2006, five years after the Thirtymile Fire, crew boss Ellreese Daniels was indicted in federal district court in Spokane. He faced eleven felony charges, including four counts of involuntary manslaughter. The case marked the first of its kind; never before had a firefighter been charged with negligence on the fireline.

Eventually, the eleven felony charges against Daniels were dropped. He pleaded guilty to two misdemeanor charges: lying about ordering the dead firefighters to a safer place and lying when he asserted that they had disregarded his order. Daniels had previously claimed that he commanded the dead personnel to leave the rock slide to which they had retreated and to join the rest of the crew on the dirt road and that they had rejected this order. Both statements turned out to be false. Judge Fred Van Sickle sentenced Daniels to ninety days of work release and three years of probation. Daniels was also ordered to submit to counseling for alcohol and substance abuse and to never fight fire again.

The rollout effects of the Thirtymile Fire are numerous and continue to be felt today. Many firefighters, upon passage of PL 107-203, decided not to pursue incident commander qualifications in the first place or to not take IC assignments if they were qualified. The risk of being held liable for events potentially out of their control was not worth it, they reasoned. Many of those who have opted to continue serving as incident commanders have purchased liability insurance in case they should ever be charged with negligence on the fireline. Informally, post-Thirtymile, firefighters are often advised to "lawyer up" and not say anything before talking to their legal counsel should an accident occur on their fires or incidents.

On the more positive side, firefighters are now more strictly held to a 2:1 work/rest ratio, meaning that if a person works sixteen hours on an incident, at least eight hours of rest must immediately follow. A new culture of safety has also emerged, with focus on leadership training and learning lessons from accidents in a non-punitive manner. The Forest Service as an agency more closely evaluates risk and whether certain fires even need to be fought. And after Thirtymile, the Forest Service replaced its old fire shelters with a newer, more effective model.

The Waldron Creek and Thirtymile Fires are separated in time by only seventy years, yet the differences between them could not be any more striking. On Waldron Creek, the "investigation" into the five deaths amounted to two minor acts: a signed statement by three of its survivors attesting that the Forest Service did all it could to keep the men safe and the county coroner assigning blame to each of the men who died. That's it. End of analysis. Investigation complete. Conversely, on the Thirtymile Fire, multiple investigations were conducted, involving hundreds of people and costing millions of dollars. A federal public law was created, a firefighter crew boss was indicted on criminal charges and Forest Service policies and practices were created or modified in myriad ways, with ramifications that persist through today. Two tragic fires, two wildly divergent reactions to them.

The most logical question to ask when addressing this dichotomy is why the differences? Why did the five victims of the Waldron Creek Fire pass into history virtually forgotten while the four deaths in Thirtymile spurred massive change? Only seventy years elapsed between the two fires, a relatively short amount of time in the grand scheme of things. Was human life less valued during the 1930s? Had the hardships of the Great Depression desensitized people to tragedy? Did folks just not really care back then when a few firefighters died?

The answers to all those questions are not clear. In fact, they may never be known. Theories and assumptions might be all we have to offer. Differences in media coverage probably account for some of the disparity. When the five died at Waldron Creek, it was days before the local papers even learned of it and ran stories about the tragedy. Television as a medium was still in its infancy in 1931. The first regularly scheduled TV news broadcast was nearly ten years in the future. Outside a fairly confined geographic area, few people even heard about the firefighter deaths in Waldron Creek. Conversely, when the four firefighters lost their lives in Thirtymile in 2001, the Internet and multiple twenty-four-hour cable news networks provided nearly instantaneous and nonstop coverage of the event. The public could not help but notice.

In Waldron Creek, the Forest Service tried to wash its hands of the fire as quickly as possible, and it did so with amazing speed and success. Within nine days after the blaze, Ranger Streeter had gotten the signatures he needed from the three survivors on his statement attesting that the agency had done all it could to protect the five men, but they died anyway. After this document was submitted, the file on Waldron Creek was apparently closed. When I say "file," I mean that in the figurative sense since no physical file of records about the fire appears to exist.

In an effort to learn more about the Waldron Creek Fire, in June 2007 I filed a Freedom of Information Act (FOIA) request with the Forest Service for any records it might have on the incident. My reasoning was that, since five men had died while being employed by the organization, surely it must possess some reports about the fire. Less than two weeks later, I received a response from the Region 1 regional forester Thomas Tidwell, who would later go on to be chief of the entire Forest Service. The reply letter was brief, but one section in particular stood out. It read, "A search for records was conducted at the National Interagency Fire Center, Northern Region, and Lewis and Clark National Forest and no records were located that are responsive to your request." Near as I have been able to determine, the next time the topic of the Waldron Creek Fire was raised in a public manner was when Nancy Thornton did so in her *Choteau Acantha* story of 2003, four years before I filed my FOIA request.

The Forest Service as an organization seemed much less tolerant of firefighter deaths when the four died on Thirtymile. It actively and comprehensively investigated the disaster and created or modified policies to lessen the likelihood that such a catastrophe could occur again. Whether this was due to its fear of the public's response should it not act (and the reactions of the politicians who represented it) or because as an organization it had evolved to the point where it cared much more about the safety and well-being of its workers is probably open to debate. I would like to believe that the latter is more accurate.

The potential for legal action also probably played some role in the differences between investigatory responses to the two fires. In 1931, it was virtually unheard of to file or pursue a legal claim against any entity of the United States government. With really no threat of being held liable should firefighter deaths occur, the Forest Service was much less compelled to fully examine itself when fatalities did happen. That had all changed by 2001. The litigious American society was in full force by this time. Lawsuit after lawsuit was filed after Thirtymile, against the Forest Service, its employees and the various manufacturers of the fire shelters and other personal protective equipment (PPE) that was used by the dead firefighters.

In 2006, families of the four deceased reached a confidential settlement with the fire shelter manufacturers and the National Association of State Foresters for defective design and faulty use instructions. One year later, Bruce and Paula Hagemeyer, the two campers who were saved by firefighter Rebecca Welch when she harbored them under her shelter, sued the Forest Service. The Hagemeyers claimed that Incident Commander Daniels took no actions to protect them as they waited for the fire to arrive. Despite the fact that one of the firefighters did in fact shield them, and that she did so at extreme risk to herself, the Hagemeyers pressed forward with their suit. They were eventually awarded a $400,000 settlement.

Also in 2006, Ken and Barbara Weaver, parents of deceased firefighter Devin Weaver, filed suit in the U.S. District Court in Yakima, accusing Forest Service employees Maureen Hanson and George Jackson of violating the civil rights of their son by destroying or withholding crucial evidence. The Weavers argued that Hanson, a forest supervisor at the Okanogan-Wenatchee National Forest, had disposed of the remains of Devin's fire shelter in order to impede their lawsuit against the shelter manufacturers. Jackson, the other codefendant in the suit, was a staff member at the Missoula Technology and Development Center. The lawsuit contended that he withheld crucial information about the shortcomings of the shelter, which forced the Weavers to settle their claims against the manufacturers for much less than they otherwise would have. He was eventually dismissed from the case when the judge ruled that he could not be linked to the shelter's disposal.

Ken Weaver blamed the Forest Service for a "culture of invulnerability" that he cited as the reason his son lost his life. "I firmly believe that they did destroy and withhold evidence for the same reason they led Devin down a dead-end road," he commented. He promised to pursue the case "with every legal breath that I have left." Weaver added that if he failed to do so, "the next time they kill kids, it's going to partially be on my conscience, and that's not something I can live with." In April 2009, the lawsuit against Hanson was dismissed, bringing to a close all legal actions related to the Thirtymile Fire. "While our hearts go out to the families of the firefighters whose sons and daughters perished in the tragic Thirtymile Fire, I am pleased that we were able to successfully defend the actions of a conscientious public servant and vindicate her reputation," stated Thomas Moss, the U.S. attorney in Boise, Idaho, who handled the civil litigation.

The disparities in the reactions to the two fires are conspicuous. Yet similarities exist as well—one just has to look more closely to find them. On both blazes, to varying degrees, the firefighters themselves were blamed for their own deaths. In Waldron Creek, Coroner Roberts explicitly faulted the dead men when he noted, "No one to blame but himself." In Thirtymile, the blame was not nearly as blunt, but it was still there. Of the fourteen significant causal factors identified in the Forest Service's investigation report, four placed the responsibility directly at the feet of those killed, including a failure to identify appropriate escape routes and safety zones, a failure to prepare the deployment site, the selection of the deployment site itself and improper use of personal protective equipment.

Granted, every firefighter, regardless of in which time period he or she worked, is accountable for his or her own actions. If things go wrong, even fatally wrong, firefighters must be willing to shoulder some of the responsibility if they themselves contributed to the accident in some way. In the Thirtymile report, while a number of the actions of those who died are listed as causal factors, also cited are management actions/inactions, as well as environmental factors like low humidity, high temperatures and dense fuel loads. However, to completely lay the blame on the dead firefighters and leave it at that, like was done in Waldron Creek, misses the mark. It's too easy to fault the deceased for their screw-ups. They are not around anymore to defend themselves and their actions. And it fails to consider other factors that might have contributed to the tragedy.

Fatal fires need to be analyzed carefully. This did not happen in Waldron Creek, nor did it take place with dozens of other deadly fires before and even after Waldron burned. When this analysis does not occur, the same mistakes run the risk of being repeated again and again. Instead of simply denouncing the dead and chalking their deaths up to human mistake, the goal of such an examination should be to realize that error can be the result of flaws in a system as opposed to flaws in character.[27] Do the five men who died in Waldron have some accountability in their own deaths? Yes they do. For whatever reasons, they chose to head in the direction of the main fire as it swept toward them. They consciously decided to leave the others. But what were the other flaws in the system that also contributed to their demise?

Like in Thirtymile, fatigue must have been a factor. Consider what the day of August 25, 1931, looked like for the men who died. Up before dawn, they had been loaded onto train cars for the ride to Choteau and then driven thirty miles or so over back roads to the fire camp, followed by a grueling twelve-mile hike to the fire ground carrying all of their equipment and tools. Chances are these men were not in prime physical condition. By the time they reached the spot fire, most—if not all of them—had to be feeling the effects of what had already been a very long day. Fatigue not only affects our physical capacity to perform, but it also has an impact on our mental functioning by challenging our ability to make effective decisions.

Communication, or lack thereof, must have also played a role. Due to the absence of portable radios, Fellers, the guide of the twenty recruits, was unable to talk with the leaders at the main fire. Thus, when the crew reached the spot fire, Fellers needed to leave the men and go converse with alternate ranger McNeal to let him know they had arrived. Without radios, communications were hampered. All exchanges of orders and information had to happen face to face, and this took time. Unfortunately, time was a commodity of which the doomed men had very little to spare.

With Fellers gone, this left the twenty firefighters without any form of supervision. No one was in charge that we know of. There was no chain of command to follow because no chain of command had been established. When the proverbial shit hit the fan and the fire started blowing up, Gunnarson apparently stepped in to fill this leadership vacuum. It would be easy to blame him for leading the four others to their deaths. But that would be a mistake, as well, since too many details are still unknown.

The Forest Service specifically—and the federal government as a whole, for that matter—must also bear some of the responsibility for the Waldron Creek deaths. Flaws in the system could be traced back to that level. Because the service hired a workforce from bars, soup kitchens and unemployment offices and then put these hires into a high-risk work environment (wildland fire), it is not surprising that negative outcomes occurred. Compounding this was the fact that the government provided these men no formalized fire training, safety equipment or defined leadership structure. It's a wonder more weren't killed under such a system. Since the government had a fair degree of culpability for the fatalities, it should have stepped up and, at a bare minimum, provided headstones and some sort of survivor benefits for the relatives of the deceased. Unfortunately, this did not occur.

When comparing fires of the past with those of today, another possible similarity emerges: the valuation of resources over human life. The men in Waldron Creek died trying to save trees. The timber, as has been argued previously in this book, was perhaps seen as being more important than the men trying to save it. It had previously been sold to a local logger, and the Forest Service felt responsible for keeping it protected until it could be harvested. An argument could be made that when it comes to fire suppression today, the houses and cabins of modern life have become the lumber of yesteryear.

Across the United States, there are now forty-six million homes on what fire professionals call the Wildland Urban Interface (WUI), and eight million more are expected to be built over the next ten years.[28] The Wildland Urban Interface comprises those areas where homes are built among or near lands prone to wildland fire. By some estimates, nearly one-third of all Americans now live in the WUI.[29] Wildland firefighters have died trying to save houses. Unlike their structural firefighting brethren, most wildland firefighters receive little or no training in protecting buildings from fire. Yet more and more often, they find themselves working in this interface environment.

On October 26, 2006, five Forest Service fire personnel, crew members of Engine 57 based out of Idyllwild, California, were killed on the Esperanza Fire near Cabazon, California. At the time of their deaths, they were attempting to defend an unoccupied residence. Eventually, the fire burned some forty thousand acres, including thirty-four houses and twenty outbuildings. The structure the dead firefighters were trying to save was included in that list. "They loved doing their jobs," commented Jeanne Wade Evans, supervisor of the San Bernardino National Forest, "but they also loved going home afterward. This time they could not go home." Raymond Lee Oyler was found guilty of arson for starting the fire, and he was also convicted of first-degree murder in the deaths of the five firefighters. He was sentenced to death in 2009. Oyler currently sits on death row in San Quentin State Prison.

In 1994, fourteen firefighters were killed on the South Canyon Fire near Glenwood Springs, Colorado. Residents of Canyon Creek Estates, an affluent subdivision of homes near the fire, became increasingly concerned about the blaze. This homeowner apprehension prompted local authorities to take action. In addition to a local crew assigned to the fire, two loads of smokejumpers, a Hotshot crew and a helitack crew were eventually assigned to the blaze. Nine hotshots, three jumpers and two helitackers were subsequently killed when the fire raced uphill, overtaking them as they attempted to flee. According to Wayne Williams, a smokejumper who was at South Canyon, the placement of firefighters between homes and fires "happens all the time."

More recently, the Yarnell Hill Fire near Yarnell, Arizona, claimed the lives of nineteen members of the Prescott Fire Department's Granite Mountain Hotshots on June 30, 2013. Only one of the hotshots survived. He was serving as a lookout for the crew from a different vantage point and was not with the rest of them when tragedy struck. It stands as this country's deadliest wildland fire since the Griffith Park Fire in 1933. Twenty-nine were killed in that Southern California blaze.

For reasons still unknown, the Granite Mountain crew left the safety of an already burned area—what firefighters call "the black"—and attempted to walk 1.6 miles through mostly unburned brush to another safety zone. That location was the Boulder Springs Ranch. They never made it. A wind shift sent the fire racing toward the crew members, and they were unable to get out of its path. Despite deploying their fire shelters, exposure to two-thousand-degree temperatures and asphyxiating smoke proved fatal for all nineteen of them.

Some in the fire community have hypothesized that the Granite Mountain Hotshots were trying to save the ranch. Others, such as John Dougherty from Investigative Media, have speculated that the reason the hotshots moved is because a state fire supervisor asked them if they could assist with evacuation efforts in the town of Yarnell. That request apparently happened just minutes before they left the confines of the safety zone. Yarnell is located just to the east of Boulder Springs Ranch, in line with their final direction of travel.

Multiple investigations of the Yarnell Hill Fire were conducted. One of those was by the Arizona Division of Occupational Safety and Health (ADOSH). The report compiled by the ADOSH cited a number of critical mistakes by the Arizona State Forestry Division (ASFD), the agency in charge of the fire. One of the findings noted that the ASFD placed an emphasis on protecting structures ahead of firefighter safety, which led to the deployment of firefighters into dangerous situations to attempt to protect property that was "indefensible." Sadly, we will never know the true reason, or reasons, why these nineteen young men died. But the possibility exists that, once again, just as it did in Waldron Creek, the value of human lives took a backseat to saving a resource.

Much has changed since the Waldron Creek Fire of 1931. Today, thousands of federal wildland firefighters are available for duty, employed by the Forest Service, the Bureau of Land Management, the National Park Service, the U.S. Fish and Wildlife Service and the Bureau of Indian Affairs. Thousands more are hired by state, local and tribal governments, in addition to privately contracted firefighters, volunteer fire departments and even inmate crews. All must undergo training in basic fire behavior and study how human factors such as leadership, communication and team dynamics affect their performance. Firefighters must pass a Work Capacity Test to confirm they are ready for the physical rigors of the job. They are outfitted with hardhats, flame-resistant Nomex pants and shirts and even earplugs and safety glasses for their personal protection. There exists a complex qualification system to ensure that those in key positions have the necessary knowledge, skills and abilities to effectively and safely do the job. Should a firefighter be killed in the line of duty, extensive analysis and investigation occur to determine how it happened and why. And numerous avenues of support are available to firefighters and their surviving family members when injuries or fatalities occur. None of these was in place when Waldron burned.

Alas, much, too, has remained the same. Fires continue to burn, whether they are the result of human causes or natural forces. People continue to struggle with their relationship to fire. Although we are finally realizing that fire cannot be excluded from the ecosystem, we complain about the smoke when fire impacts us directly or demand that blazes be put out when they burn in our proverbial backyards. Firefighters continue to die in the line of duty. The men in Waldron Creek perished when fire got below them in a tight, fuel-filled drainage that was subjected to high, upslope winds. Those same circumstances would occur again with deadly results in Mann Gulch (1949), South Canyon (1994) and the Cramer Fires (2003). Regrettably, they will probably align again in the future, killing more fire personnel. Lastly, just as happened in Waldron Creek, firefighters still die trying to save resources, be it homes or timber.

No one seems to be overly shocked when a policeman is killed on the job in a vehicular accident or shootout. Likewise, we are not too stunned when structural firefighters are killed in a building collapse or a flashover. Saddened, yes, but not surprised. It goes with the territory. For some reason, though, people seem to be caught off guard when wildland firefighters die in the line of duty. Yet wildland firefighting is a high-risk occupation, just like law enforcement and structural firefighting. That much will never change. It was dangerous in 1931, when our Forgotten Five died in Waldron Creek.

It continues to be hazardous today, as highlighted by the nineteen deaths on the Yarnell Fire in 2013.

So much has changed since the deaths of the Forgotten Five, and yet so much remains the same. Wildland firefighters have died in the line of duty—almost one thousand of them, in fact, since the Waldron Creek Fire in 1931. The sad reality is that even more will die in the future. We lament when they are killed trying to save trees and houses, yet ironically those are the things we hire them to protect. If we continue to expect that firefighters will place themselves in harm's way to defend us and what we hold dear, then we need to come to grips with the fact that some will die doing just this. Therefore, we owe it to firefighters, and to ourselves, to support them to the highest degree, with the best training, equipment and budgets, so that the number of those killed is kept as low as possible.

5

FIRST STEPS

As noted in the previous chapter, my FOIA request to the Forest Service in 2007 for records on the Waldron Creek Fire yielded nothing. If the agency did not possess any documents on the blaze, there were only a couple reasons why this would be the case. Either archives of the fire had been unintentionally or purposefully "misfiled" at some point and therefore effectively lost, or there actually were no records on the fire. Either way, the end result for me was the same: I was not going to find out much, if anything, about the fire from the Forest Service. A quote from *Young Men and Fire* seemed to be on a continuous loop in my head: "To the average citizen the government holds nearly all the cards and will play them when the government is both the alleged guilty party and judge of its own guilt."[30]

Truthfully, I had pretty much expected that my FOIA request would not be very productive. In 1993, the National Wildfire Coordinating Group, of which the Forest Service is a sponsor, first published a document titled *Historical Wildland Firefighter Fatalities (1910–1993).* Although the Waldron Creek Fire clearly falls within that range of years, no record of it could be found within the publication. The only fatality fire registered for 1931 was an incident called the Coffee Mill Fire near Mariposa, California, where three men were killed when overrun by flames. Since Waldron Creek was not listed in the book, I interpreted this to mean the Forest Service did not have any institutional memory of the event.

Having failed to find anything with my FOIA efforts, I was at somewhat of an impasse. More and more, I began to feel that if I could just get something in place that was tangible evidence of their deaths, it would be easier to gain some momentum on this project. One call to the Wildland Firefighter Foundation (WFF) in Boise, Idaho, and the wheels of progress began to turn very quickly. I was put in contact with Vicki Minor, the founder and director of the WFF. For more than twenty years, Vicki ran her own business, selling dry goods to firefighters at camps throughout the West. After the South Canyon Fire in Colorado in 1994, which claimed the lives of fourteen firefighters, she realized there was a tremendous need, unmet at the time, to provide assistance to the families left behind following such a tragedy. Thus was born the WFF.

Its mission remains the support of family members of those injured or killed in the line of duty. To date, thousands of people, many of them children who have lost a parent, have been assisted. The majority of the WFF's funding comes from its own rank and file: firefighters themselves. Through the slogan of "a buck a week to help a buddy," the WFF gets firefighters (thousands of them, at last count) to contribute to the "52 Club." Besides financial help to families, assistance from the WFF can and does come in other forms: immediate and ongoing emotional support, advocacy and recognition to fallen and injured wildland firefighters. It was the latter of this list, the recognition piece, that I was the most interested in exploring.

It became Vicki's dream to do something to tangibly honor those who have fallen, and through her efforts and the contributions from a multitude of volunteers, her vision became a reality. The Wildland Firefighter Monument in Boise helps to pay tribute to those firefighters who have perished in the line of duty or who, during their lives, provided significant support to the wildland fire community. What was once a parking lot at the National Interagency Fire Center was transformed into the "Wildland Firefighter's Sanctuary" that it is today. After five years from conception to completion, the monument was officially dedicated on May 25, 2000.

On that day, scores of people associated with the wildland fire community across the nation came to witness the proceedings. A C-130 retardant plane and a DC-3 smokejumper aircraft did flyovers. When overhead, they dropped not their usual payloads of "mud" and men but instead released a cluster of purple streamers, which silently drifted from sky to earth, alighting amid the spring wildflowers that had taken hold at the monument site.

At the core of the monument stand three eight-foot bronze statues of firefighters, created by sculptor Larry Nowlan, who began working on the figures while he was an artist-in-residence at Saint-Gaudens National Historic Site in New Hampshire. According to writer and fellow artist

Sculptures at the Wildland Firefighter Memorial, Boise, Idaho. *Author's collection.*

Michael Horvath, "[Nowlan's] imagery is accessible and beautifully crafted, yet embodies the hidden qualms and hopes of our existence with genuine dignity. That which is us—that which we can be—is to be found in his work." The oversized dimensions of the sculptures are not an accident or oversight; they are meant to convey the larger than life and heroic status that firefighters hold in the public consciousness.

The monument typifies so much about the fire community. People came together simply because they realized doing so was the right thing. Hundreds and hundreds of individuals donated their time, energy and talents without concern for recognition or compensation to make sure their fallen comrades would not be forgotten, that the place where they will be remembered is fitting and worthy of their sacrifice. The National Wildland Firefighter's Monument is surely that and a great deal more.

Every year since 2005, the WFF has hosted a Family Fire Day in May, when family members come to the Foundation in Boise to recognize and remember their fallen firefighters, to do some healing and to meet other survivors. According to Vicki, "The amount of healing and friendship building that occurs for the families is amazing. They really come away feeling like they are not alone, they have someone to turn to, and there is a real community and bigger family they are a part of." A central part of the Family Fire Day is the unveiling of that year's markers at the monument, which honor firefighters who have died in the previous twelve months. Luckily for the men of Waldron Creek, the foundation was willing to make an exception to this time frame. Although they died in 1931, the staff of the WFF was receptive to placing a marker to honor their sacrifice during the May 2009 ceremonies.

Finding them willing to do this, I quickly began making plans to attend the Family Fire Day event. Talking with one of the staff at the WFF, I discovered the dates for the ceremonies would be May 16 and 17. As luck would have it, of course, my schedule made attending quite a challenge. I had already committed to serving as an official for the Big Sky Conference Track Meet on Friday, May 15, and for the morning of Saturday, May 16. On Saturday afternoon, I was likewise committed to graduation ceremonies at the University of Montana. If I was going to be able to attend, it would be just for Sunday, and it would mean leaving Missoula on Saturday evening for the roughly eight-plus-hour drive to Boise.

Family Fire Day: May 2009

One might question the logic of driving a total of more than sixteen hours to attend a three-hour event, especially when that drive begins late in the day and takes a person through some of the windiest-ass roads the West has to offer. To me, it was a no-brainer. I had to go. After the graduation ceremonies, I hurried home to pack the few things I needed and to get at least a little time with Christine and the kids before hitting the trail. My plan was to leave town by 6:00 p.m., which would put me in Boise by about 2:00 a.m. the following morning. Amazingly, I was damn close to this departure time, on my way out of town by about 6:30 p.m. I stopped at the Conoco in Lolo to fill up on fuel and junk food before hitting Highway 12 for the curvy trip over Lolo Pass and into the Lochsa country.

I nearly made it to Lolo Hot Springs, about twenty-five miles, before I realized I had forgotten in my office a very important file that I needed for the ceremonies. For a few miles, I toyed with the idea that I would not need it, but my preoccupation with always being prepared eventually won out. Quickly, I flipped a U-turn and headed back the way that I had come, beating myself up the entire time for being such a dumb shit. A long trip had just been made even longer due to my mistake. It was not a great start.

I managed to hang on until about 2:30 a.m., which is about the time I hit Interstate 84 for the final fifty-mile push into Boise. When I saw a large white rabbit and what appeared to be Sasquatch hitchhiking on the side of the road, I knew it was time to shut things down. Pulling into a truck stop, I drove around with my lights off until I found a little spot where I could nestle in between two rumbling semis. I had my Thermarest and a thin sleeping bag, which is all I really needed considering I was going to get only a few hours of rest in the back of my car anyway. I didn't even bother taking off my clothes. With the smooth idling of the nearby big rigs, I was out within seconds.

What seemed like only minutes later, the alarm on my watch began to chime, forcing me out of my sleep coma. Looking around, I noticed that almost all of the semis had left. I hadn't even heard them go. I pulled my shoes on and headed for the truck stop diner. I looked over the menu, made my order and then casually glanced at the old Pepsi clock that was hanging on the wall. I did a double and then a triple take. According to the clock, it was an hour later than I thought. Before racking out, I had set my watch back one hour, assuming I was on Pacific Time. At the top of Lolo Pass when you enter Idaho, the time zone changes. Unfortunately, in my driving blur, I missed the highway sign notifying me that once I crossed the Salmon River,

I was leaving the Pacific zone and heading back into Mountain Daylight time. My little snafu left me in a rush. So much for being ahead of the time curve, I thought. Now I had barely an hour to eat and finish the remaining drive into Boise.

While scarfing down the nasty blueberry pancakes I had ordered and mainlining some really bad coffee, I was forced into making small talk with a heavy-set trucker from Florida who was sitting next to me. After minutes of him telling me about how hard it was to make a living at a dollar a mile and how the economy is going "in the shitter," he asked me what I was hauling. Being the smartass that I am, I contemplated telling him, "Three empty Coke bottles and a can of Copenhagen," but I resisted. OK, the jury has spoken. I must look pretty haggard if I'm being mistaken for a truck driver. Without meaning to offend my trucker brethren, they generally aren't the cleanest, most well-groomed population. I broke the conversation short, threw some money on the counter and headed for the restroom. A quick washing of my hair in the sink with some hand soap and I was out of there, ready for the final leg of what had been an already long trip.

Luckily, the remainder of the drive into Boise was uneventful. I arrived at the WFF about twenty minutes before things were set to kick off. I checked in at the reception desk, made small talk with a few of the other folks who were also walking around and then parked myself at an empty table. Fairly quickly, my table and all of the others began to fill. Since this was the second day of activities, most everyone else seemed comfortable and at ease with the proceedings. As it turned out, the people who sat down next to me were family members of Andy Palmer. Andy had been killed in July 2008, when he was fatally struck by a falling tree on the Eagle Fire in the Shasta Trinity National Forest in California. The Palmer family was in attendance for the unveiling of Andy's memorial marker at the monument. Together, we joked about the coincidence of a group of Palmers ending up at the same table. Silently, I wondered if it was merely happenstance.

Moments later, Vicki approached the podium. She thanked everyone for coming, and then she went over the list of events for the day. A motorcycle honor guard was set to kick things off, complete with dozens of Harley Davidson riders carrying American flags. Other activities included sign making and a release of balloons to honor the fallen. Lastly, white doves were to be brought forward. Friends and family members of the deceased would have the opportunity to touch the small birds before they were set free. It was a moving rite. The rest of the day, other than time for a few

Right: A white dove just prior to release at the Wildland Firefighter Foundation Family Fire Day, May 2009. Boise, Idaho. *Author's collection.*

Below: White doves in flight. Wildland Firefighter Foundation Family Day, May 2009. Boise, Idaho. *Author's collection.*

speakers, was set aside for families and friends to tour the grounds and to interact with and support one another.

One of the few speakers that day was the mother of Rob Browning. It soon became apparent why she had been chosen to talk. She was a captivating presenter. Rob Browning and Rich Tyler were the two helitack crew members who died in the South Canyon Fire in 1994. While the other twelve who died in the fire were caught and overrun on what has been called the West Flank fireline, Browning and Tyler's race against death had originated at H2, or Helispot 2. This position was upslope and east of the West Flank line, along what has been dubbed Hell's Gate Ridge. Attempting to get to higher ground for a possible helicopter pickup, the two were burned over approximately one-third of a mile from H2. Despite the differences in locations between the two groups, the end was, tragically, the same.

Mrs. Browning talked about how, for eighteen months after her son's death, she struggled to cope with the loss. Finally, she decided she needed to see the actual place where Rob had died in an attempt to bring some semblance of closure. All the way up the arduous hike from the trailhead, she and her husband prayed for some sort of indicator to let her know that Rob was in a better place. "Please, Lord, show me a sign, anything—a white butterfly perhaps," she pleaded. Shortly after she uttered her prayer, a snow-colored butterfly appeared, even going so far as to land on her husband's hand as they rested from the grueling hike. "I got my wife back that day," shouted Mr. Browning, who was in the crowd as his wife spoke.

To this day, I still find it hard to comprehend what happened next. Seconds after recounting this in her speech, a white butterfly hovered over the crowd, hanging for a few tantalizing moments before floating away. For an instant, I thought I must be seeing things, but then the assembled throng began to murmur. As I looked around, I realized others had seen it, too, and its symbolism was not lost on those in attendance. "It's a miracle," I heard someone exclaim. At that moment, I found it difficult to disagree.

All my life I have struggled with religion and the age-old question of whether God really exists. My parents made it a point to bring me and my brothers to church fairly regularly when we were growing up. For whatever reason, I have not done the same with my own children. To this day, I have strong memories of sitting in those church pews as a kid, fidgeting, waiting for the service to be over. I was not ready for religion then. I'm not sure if I'm ready for it now. I want to believe, but I also want some evidence. Faith with conditions. I know it doesn't work this way, but that is where I have been stuck. Perhaps part of the proof for which I have been searching came

The memorial marker for victims of Waldron Creek Fire, Wildland Firefighter Memorial, Boise, Idaho. May 2009. *Author's collection.*

in the shape of a white butterfly. Sure, simple coincidence could account for its appearance on these two different occasions, but the probability of that happening has got to be quite low.

With most of the events and speakers having winded down, I wandered through the monument looking for the marker for the Waldron Creek Fire. I had resisted searching for it up until that point. I wanted to fully experience the ceremonies first and let my excitement build, since this marker was set to be the first real, tangible symbol paying tribute to the Forgotten Five. I meandered for several minutes, looking, but without success. The freshly disturbed soil caught my attention. There it was. I hesitated, and then I moved forward for a closer look.

I noticed the errors right away. I couldn't help myself. I had so wanted this moment to be perfect, and now it was anything but. Months earlier, I had sent the foundation information about the incident: the names of the men who died, the name of the fire, the place where the fire occurred, the date of their deaths and, lastly, the epitaph: "We Will Remember." Somewhere in the process of the foundation providing these facts to the engraver, there had been a breakdown in communication. Where it should read "NF," as in National Forest, it instead says "NS." The other error was much more significant. The year was wrong. The fire occurred in 1931, not 1935 as engraved.

My initial impulse was to grab someone from the WFF and point out the blunders. "This is wrong!" I wanted to scream. I was too close, too invested, which left room for emotion to override common sense. Thankfully, I resisted that urge. I took a deep breath and tried to slow down my mind. There would be a time and place for bringing it to their attention, but that time and place was not now. Flight, not fight, was the best course of action at the moment. The most appropriate thing to do was leave before I did or said something I would later regret. I quietly slipped away from the ceremonies, hit the road and commenced to stew on it the entire way back to Missoula. Despite all the powerful and positive events of Family Fire Day 2009 at the Wildland Firefighter Monument, I was left conflicted and torn. This was not how I had pictured the day would be.

I gave myself twenty-four hours to process what had transpired. The conclusion I reached was that inaction was not an option. I had to say something about it. Sure, errors can and do happen—after all, many mistakes had contributed to these five men being killed—but in this case, the mistakes could be corrected. I called the staff member at the WFF with whom I had worked previously when trying to get the marker in place. I told her about the two issues on the marker. She sounded appreciative that I was bringing it to her attention but aloof at the same time. "I will look into it," was how the conversation ended.

And then, I dropped it. I never followed up to see if anything had been done to correct the errors. There were plenty of other things to do with this project, so I shifted my attention to them. Before I even realized it, five years had passed. While my travels had brought me to Boise in a couple of instances during this period of time, I never visited the monument to check on the Waldron marker. Finally, in late 2014, when reworking this section of the book, I called my friend Gary Luck, who lives in Boise. I asked him if he would swing by the monument and check. Gary was out of town but promised that he would do so when he got back to Boise. A couple days later, he called back. "It's still the same. No change," was his report. Again, I contemplated what to do. Although torn, I decided to drop the matter. While the marker is not perfect, it is the first tangible step taken to remember the Forgotten Five. There is a lesson in there somewhere, I suppose, but for now I'm not sure exactly what it is.

6

HIGHLAND CEMETERY

I have always been drawn to cemeteries. It must have something to do with my preoccupation with human mortality, I suppose. After all, there is no better setting to appreciate life than in a place where you are completely surrounded by death. Death can be such a nebulous construct, so hard to understand and comprehend. Yet in a cemetery, death is right in front of your face, silently screaming at you that this is your eventual fate as well. What better tangible symbol is there of human mortality than hundreds of gravestones?

In my meanderings through various cemeteries, I have become enthralled with headstones. They must, in a few short words, tell the complex story of somebody's life. A name, dates of birth and death perhaps and maybe a brief epitaph are all that remain, yet there is so much more to their personal story. A baby who died shortly after birth, a soldier killed in war, a woman who lived to be one hundred—these and so many more.

Highland Cemetery in Great Falls, Montana, like all cemeteries big or small, is an intriguing place. But then, to me, all cemeteries are extremely engaging. I mean, the collective history that is housed in all of those dead bodies is mind-boggling. When I think of cemeteries, though, it's hard to forget the (in)famous line from Rodney Dangerfield in *Caddyshack* when he, as the character Al Czervik, opines that golf courses and cemeteries are the biggest wastes of prime real estate. Rest easy on this one, Rodney. It would be a stretch to say the acreage that composes Highland Cemetery is "prime real estate." Located on the far southern fringes of town, not much

Entrance to Highland Cemetery, Great Falls, Montana. *Author's collection.*

surrounds the place, save for some grain fields and a few scattered houses. From an aerial viewpoint, Highland stands out as an island oasis of green trees and grass in an otherwise large ocean of brown.

The cemetery was established in 1911 and quickly found a steady supply of customers as World War I escalated. Today, that portion of the cemetery is simply referred to as Old Highland. New Highland lies just to the northwest, and it is here where two of our Waldron victims rest. Like most cemeteries, it is rumored to be haunted, although sightings of anything strange or ethereal are infrequent at best. One Highland worker, since retired, swore to seeing a horse-mounted ghost rider on the grounds, but he was never able to catch the spirit's name or find out just why he was so restive.

Upon entering the cemetery through its impressive stone-and-iron front gate, the first monument you encounter is that of Paris Gibson, the "founder" of Great Falls. Born in Maine in 1830, Gibson kept moving west as he grew older. After a stint on the University of Minnesota Board of Regents from 1871 to 1879, he left the Land of Lakes after his business interests there failed to produce. In 1880, he paid a visit to the Great Falls of the Missouri. With his keen eye for the almighty dollar, he realized the economic potential

The memorial marker for Paris Gibson. Highland Cemetery, Great Falls, Montana. *Author's collection.*

of harnessing the falls for the production of hydroelectric power. Gibson just happened to be a close friend of railroad magnate James J. Hill, and he managed to convince Hill that it would be in both their best interests to develop a town site at the falls. With Gibson's business plan, and Hill's money, Great Falls was born.

Of course, others had been here well before Gibson first laid eyes on the place, but he is considered the "founder" since none of the others had felt like stopping for too long. Paleo-Indians first migrated into the area almost twelve thousand years ago, weaving their way down the eastern side of the continental divide, between the retreating ice sheets. Thousands of years later, Salish Indians from what is now western Montana hunted bison here on a seasonal basis but established no permanent settlements. By the early 1600s, Piegan Blackfeet (the largest of the three tribes of the Blackfeet Confederacy) had entered the area, driven the Salish back into the Rockies and declared the land their own.

The most notable of the previous visitors would have to be Meriwether Lewis, William Clark and their Corps of Discovery, who spent one month

The Great Falls of the Missouri River, date unknown. *University of Montana Archives and Special Collections.*

here portaging around the falls between the middle of June and the middle of July in 1805 on their way to the Pacific. Harry Fritz, a now retired professor of history at the University of Montana, once quipped, "Thirty days in Great Falls, I wouldn't wish that on anyone." Needless to say, that comment didn't play too well in the Electric City, but it did bring chuckles to others, many of whom do not really consider "G-Funk" to be the most interesting of places. A year later, on the return trip, Lewis and three other men from the corps would skirmish with a small group of Piegans just downstream from the falls, resulting in the stabbing death of one Indian and a presumably fatal wound to another. Lewis and his band made a hasty retreat, covering over one hundred miles on horseback in a single day, before rendezvousing with others from the corps farther down the Missouri River.

I have always been fairly conflicted with regard to how I feel about Great Falls. Since it is my hometown, I think there is some sort of psychological hardwiring that predisposes me to not like the place. Either that or the constant wind blew through my ears hard enough as a child to dry out my brain (Great Falls is consistently ranked as one of the windiest places in the United States). I mean, kids just sort of grow up hating the place they are from, don't they? That is probably what helps to push you out of the nest eventually, causing you to seek out other locales and opportunities. As a youngster growing up there, Great Falls just seemed so dull, and that is a feeling many people continue to have about it. It just doesn't seem to have the vibrancy of cities like Missoula and Bozeman. But as I have grown older,

I've moved past judging Great Falls by what it lacks and have learned to appreciate it for what it does have: my parents and brother still living there, engaging citizens, wide-open places, some cool old buildings and quick access to great fishing and hunting spots. I actually look forward to making trips to Great Falls now, and that is most certainly a feeling I never thought I would possess during my younger years.

Also buried at Highland is one George Montgomery Letz—or at least half of him. The other half resides in the Forest Lawn Cemetery near his home in Rancho Mirage, California. The youngest of fifteen children, Letz was born in Brady, Montana, and grew up on the family ranch there, where he learned many of the skills he would later use as one of Hollywood's leading cowboy actors. Letz had attended the University of Montana for a year, but in 1935 he dropped out and headed to Los Angeles to find work in the movies. As a professor at UM, I love the irony that oozes from the fact that one of our most famous students is a dropout. It did not take the ruggedly handsome Letz long to find work, as Republic Pictures hired the then nineteen-year-old to perform stunts. He continued to do stunt work and play small parts in various movies and even appeared in the *Lone Ranger* serial picture in 1938.

Shortly thereafter, he signed on with rival 20th Century Fox studio but did so as George Montgomery, the Letz name having now been dropped. He went on to leading man roles in many pictures, including *Coney Island*, *Ten Gentlemen from West Point* and *China Girl*. In 1943, the same year he married singer Dinah Shore, he enlisted in the U.S. Army Air Corps and served three years before resuming his acting career in a long string of movies and television shows. Montgomery and Shore divorced in 1963, the same year Montgomery's obsessive-compulsive housekeeper attempted to kill him, and then herself (she failed in both accounts), because she thought he spent too much time womanizing. Montgomery continued his work in motion pictures into the 1980s and was also an accomplished painter, sculptor and carpenter. He died in December 2000.

Perhaps the most famous "resident" of Highland Cemetery is Charles Marion Russell, aka C.M. Russell. Born in St. Louis in 1864, Russell dreamed of living the cowboy life. By his sixteenth birthday, he had moved to Montana's Judith Basin country and taken up work as a sheepherder. He quickly realized that tending woolies was not the most fascinating of occupations, and he eventually found a job as a wrangler with the Judith Basin Roundup. Russell spent eleven years cowboying before retiring to become a full-time artist.

George Montgomery's grave. Highland Cemetery, Great Falls, Montana. *Author's collection.*

During his lifetime, Charlie Russell completed roughly four thousand works of art, including drawings, paintings and sculptures. He died in 1926, and on the day of his funeral, all the schoolchildren in Great Falls were dismissed from class to watch the processional. Russell's coffin, displayed in a glass-sided coach, was pulled by four black horses. A riderless horse, complete with boots in the stirrups backward, accompanied the hearse. It was, and continues to be, the most attended funeral processional in the history of Montana.

Charles M. Russell's grave. Highland Cemetery, Great Falls, Montana. *Author's collection.*

Besides the more well-known individuals who are interred here, two of the victims of the Waldron Creek Fire also reside within the confines of Highland Cemetery. Frank Williamson and Herbert Novotny, best friends both in need of work when the call for firefighters was made, lie within feet of each other. Novotny was African American and Williamson Caucasian, yet the two men were able to forge a relationship that transcended the color of their skin, despite the fact that racism and segregation were being actively practiced even in Great Falls. Both men can be found in the southeastern section of the cemetery, in an area simply referred to as 10D.

The vast majority of the grave markers in 10D are of the flush-to-the-ground variety, reflecting the fact that most of these are single graves, unlike the family plots in other portions of the cemetery, which tend to have much larger tombstones and monuments. Not everything that is in the ground here is dead—gopher holes pockmark the area like some sort of landscaping acne. It seems weird to think that, underground, rodents are digging tunnels that wind under, over and around all those caskets and decaying bodies. My hunch is that cemetery workers everywhere have a warm spot in their hearts for Carl Spackler from *Caddyshack*.

After the fire, the victims were removed from the mountainside and brought back to Choteau. Hearses from the T.F. O'Connor Funeral Home and the Merrill Mortuary in Great Falls were dispatched there, and they transported the bodies of Novotny and Williamson back to their families in the Electric City. Services for Williamson were held at 2:00 p.m. on Monday, August 31, at the Merrill Mortuary. The Reverend Sturgeon officiated. Burial at Highland Cemetery followed later that afternoon. Since the Forest Service was unwilling to provide a headstone, the family pooled their scant resources and purchased a modest memorial for their son. It simply lists his name, years of birth and death, the words "AT REST" and, finally, etched in roughly the middle of the marker, a small evergreen tree, perhaps emblematic of his fateful time working in the woods.

Services at Highland for Herbert Novotny were held two days later, on September 2. A large contingent of his friends and family were present, including his wife, Agnes; their two daughters; his brother, Harvey; his two sisters; and a number of the congregation from the Union Bethel African Methodist Episcopal Church. Herbert's mom and dad were also in attendance. They had not seen him at all during the week leading up to his death and had no idea he had joined a firefighting crew. Again, the Forest Service was unwilling to pay for a headstone. Unfortunately, no one in Herbert's immediate circle had the financial means to purchase one for him. Thus, he was buried in an unmarked grave slightly down the hill and to the south of his best friend.

Even though it is in my hometown, to the best of my memory I had never visited Highland before my first trip there in 2005. My wife, our two-year-old son and I had come to Great Falls for a summer visit from Minnesota, where we were living at the time. With the help of Bob Stewart, one of the workers at Highland, I was able to find Williamson's grave marker, as well as roughly the spot where Novotny's grave should be. I took some digital pictures of the place, including Williamson's burial site, but true to my lack of organizational skills, I failed to download them before eventually losing the camera.

My now seven-year-old son, Skye, and I revisited Highland in October 2010. We had come to Great Falls for the opening of pheasant season on Saturday the ninth. A former student of mine, Dan McFarland, had invited us up to Conrad, about an hour drive north of Great Falls, to hunt on some private land he had access to. We drove over on Friday night, spent the night at my parents' house and drove up to Conrad early the next morning. We hunted most of the day, nine of us in total, and ended up with twenty-five roosters, nearly a limit. That night, Skye and I drove back to Great Falls. Instead of getting up early and hunting another day on Sunday, we opted instead for a little more leisurely paced morning, punctuated by a visit out to Highland.

On my previous visit to the cemetery in 2005, I had noticed that the etched letters on Williamson's marker were filled with dirt and other debris, making it difficult to read. For this trip, I brought a gallon jug of water, a bristle brush and various cleansers, hoping that they would make the headstone more readable. Skye and I quickly found the grave site and set to work cleaning it the best we could. Despite copious amounts of water, Comet and elbow grease, we didn't really have the cleaning effect I wanted. While we were able to get the soil out of the engraved letters, it did not really make them that much more visible. However, we did do *some* good, and the combination of the monument being wet along with some direct sunlight did make the words and numbers a little more discernible.

Since 1931, and up until the marker was placed at the Wildland Firefighter Monument in Boise, Williamson's gravestone served as the only palpable reminder of the Waldron Creek Fire. But it is only there because his family contributed the funds to purchase it. As a father of two children myself, my thoughts often turned to the widow Agnes Novotny and her two young daughters, now fatherless. It is easy to picture them in my mind at the

The grave of Frank Williamson. Highland Cemetery, Great Falls, Montana. *Author's collection.*

cemetery on that hot September day: Agnes struggling with the finality of it all; the girls, ages one and three, blissfully unaware of what it all meant. It must have been incredibly awkward and uncomfortable to be burying their loved one in such an anonymous fashion, with no headstone to mark the site. It still baffles and angers me that the Forest Service did not step up to the plate and take care of this for the five men killed in Waldron Creek.

I promised myself I would do what I could to get a marker in place for Herbert at Highland Cemetery. I vowed to do the same for Gunnarson and Allen, who likewise rest in unmarked graves fifty-five miles to the northwest in Choteau. And lastly, I made a solemn promise to Ted Bierchen that I would attempt to find his grave and get a marker for it as well, if need be. They deserve at least that much.

7

FINDING DOROTHY

In the winter of 2007, during my first academic year of employment as a professor at the University of Montana, I applied for a small internal grant from the university's research office—to the tune of $5,000. This was the maximum amount that could be requested. My intent was to use a good portion of that grant money to pay a professional genealogist to map out the family trees of the five men who were killed, to see if any living relatives might still be around and, if so, where they might be located. Unfortunately, my research grant proposal was shot down, leaving me in a quandary about how to proceed. Eventually, I decided I would bite the bullet and try to do it myself. I paid twenty bucks to the website Ancestry.com, which gave me access to its genealogical databases for one month. For hours I plugged names into the search engine, hoping for a hit. It did not take long for me to become completely frustrated with genealogical research.

I predominantly searched for Novotnys because I knew Herbert had two daughters when he died: Jane and a younger girl for whom I had no first name. I found plenty of Jane Novotnys and even talked to several of them on the phone. However, none of them was the person I was looking for. Each time I called a name from my list, I hoped that this might be the one and that slowly I could begin to put some pieces of the puzzle together. Unfortunately, that never happened, and discouraged by my lack of success, I let my subscription to the website expire.

The following year, in 2008, I reapplied for the same internal UM grant, but this time I backed off on the requested amount and asked for only $1,200. To my surprise, the committee that reviewed the applications decided to fund my proposal. Finally, I had some seed money to begin. I contacted Bryan Lockerby, a private investigator in Great Falls, and told him about my search. Bryan had several good suggestions, one of which was to utilize what he called a "skip tracer" to find out any information. A skip tracer is a person who specializes in tracking down people, and Bryan had a colleague who did just this kind of work. Apparently, according to the folks at Wikipedia at least, skip tracing is "derived from the idiomatic expression 'to skip town,' meaning to depart (perhaps in a rush), leaving minimal clues behind to 'trace' the 'skip' to a new location." Sounds good to me. Skip away, dude.

Collectively, Bryan and I decided to have the skip tracer focus on the relatives of Herbert Novotny, since I had the most information on his family background. Unfortunately, after the first several searches, not much came back. The tracer was able to identify a few Jane Novotnys, like I had with my Ancestry.com search, but none of them turned out to be Herbert's daughter.

Fortunately, I was due for a lucky break, and when it came, the pieces of the puzzle quickly started to fall into place. Looking for more help, I called the Cascade County Clerk and Recorder's Office in Great Falls to see if they might have any information on Herbert Novotny or his wife, Agnes M. Williams Novotny. The person I talked to (who shall remain nameless, since he was probably not following policy by telling me the information) was able to look at the records and determine that Agnes had, a few years after Herbert's death, remarried a man by the name of Beatrice (Bea) Parker. Mr. Parker legally adopted Jane and her sister, a girl named Dorothy. Both children then assumed his last name of Parker, which is why my traces for the Novotny girls had been off the mark.

With the help of the folks at the Great Falls Genealogical Society, I was able to track down an obituary for Bea Parker, who died on November 6, 1974, and is buried in the Manchester Cemetery outside Great Falls. An African American born in Jackson County, Missouri, in 1888, Bea came to Great Falls in 1910 to work for the Great Northern Railway. He was a maintenance man by trade and eventually was employed by Montana Bank, Bell Telephone Company, the C.M. Russell Gallery and the First Federal Savings and Loan Association. After becoming the adoptive father of the two Novotny girls, Bea had several other children with Agnes Novotny Parker. Bea, Agnes and their now large family were very active members in the Union Bethel African Methodist Episcopal (AME)[31] Church in Great Falls.

Union Bethel AME Church, Great Falls, Montana. *Author's collection.*

It was a fairly straightforward process, with assistance from Investigator Lockerby, to backtrack from Bea's obituary and find out who all of the children were and where they were living at the time of his death. Obviously, I was most interested in Jane and Dorothy Parker, since they were the biological children of Herbert Novotny, the youngest member of those killed in the Waldron Creek Fire.

Jane, the elder of the two girls, was born on August 21, 1928. Since the Forgotten Five were killed on August 25, chances are that just a few days before dying on a Montana mountainside, Herbert had been home with his young family celebrating Jane's third birthday. Some simple digging revealed that Jane Novotny married a man by the name of Leon Harris and moved to the Denver area. Jane was an accomplished singer, and she spent time performing with the Colorado Opera. Like her biological father, Herbert, Jane possessed a strong service mindset. She served in the United States Army for a period of time, attaining the rank of Specialist 5. Jane F. Harris died on September 23, 1978, at the age of fifty. She is buried in Fort Logan National Cemetery in Denver.

As the efforts to learn more about the Parker family continued, Lockerby put me in contact with Barbara Behan, a local historian in Great Falls who had done some work with the AME Union Bethel Church there. She in turn was able to connect me with Ken Robison, another historian who had also done considerable research on the church and its members. Ken turned out to be an invaluable help, since he directly knew the children of Bea and Agnes and, more specifically, had phone numbers for many of them.

Obviously, I was most interested in talking with Dorothy since she was the only still living biological child of Herbert. Ken agreed to get in touch with her and to bring her up to speed on my project. When he got back to me with the confirmation that she was willing to talk, I could hardly contain my excitement. Nervously, I dialed her number, not quite sure what to say but figuring I would find adequate words when the time came. Once I got her on the phone, I quickly realized that Dorothy Parker was still sharp as a tack and was eager to find out more about the fire and her biological father who had died there.

Dorothy, the younger daughter, was born on June 15, 1930, making her a mere fourteen months old when Herbert was killed in Waldron Creek. Her mom, now Agnes Parker after her marriage to Bea, never talked about Herbert, so Dorothy and Jane grew up not knowing much, if anything, about their biological father. To them, Bea was their dad, and he served in that role admirably. But each of the girls retained questions about Herbert that their mom was either unwilling or unable to answer. In some way—she does not remember how—as a youngster Dorothy got her hands on an old photograph of Herbert. He was about seven years old at the time. She would look at it frequently, wondering what this father she never knew was like.

Like Herbert and her older sister, Jane, Dorothy values service to others. She joined the United States Navy in 1951. Dorothy served during the Korean War, working in an intelligence section helping to analyze aerial photographs of enemy positions. She left the navy in 1955. Dorothy eventually married William Hines, and together they had four daughters. The Hines family spent a good portion of their lives living in Derry, New Hampshire. William Hines passed away in 1978, the same year that Dorothy's sister Jane died. Dorothy moved to California to be closer to other family members who lived in that area, and recently she moved again, this time to Colorado, again to be closer to family. Despite being in her mid-eighties, Dorothy continues to make elaborate dollhouses from scratch, which she donates to worthwhile children's groups.

It seemed like nothing at all happened with regards to my investigation into the Waldron Creek Fire for a considerable amount of time, and then, all of a sudden, events started unfolding almost more quickly than I could keep up with. During the summer of 2011, I was in discussions with Malisani, Inc. in Great Falls, which was working on the headstone for Herbert Novotny. Tony, one of the brothers in the family business, and I had gone to high school together, so I immediately thought of him when I needed a grave marker. Once I decided on the wording, I sent him my approval of the final rendering, along with a check for nearly $400. The price tag for this project was starting to grow.

I decided to hold a dedication ceremony for the placement of Herbert's gravestone, so the next major order of business was to determine a possible date for this event and to see if Dorothy would be able to attend. I was relieved and excited when I talked with her on the phone and she expressed an interest in coming to Great Falls for the ceremony. Since she was in a wheelchair, she told me that it would be necessary for her daughter Alicia to come with her. I told her this would not be an issue, hung up the phone and went to work online trying to book the travel for the weekend of September 24 and 25, 2011. Minutes later, Dorothy called and told me that Alicia's fifteen-year-old daughter, Sydney, would also need to make the trip since they had no one who could watch her while they were gone. Of course, I told her "no problem" and resumed my search for what would now be three plane tickets instead of two.

I had visions that Delta Airlines or one of the other carriers that served the Great Falls market would be happy to step up and either donate some or all of these tickets or at least give them to me at an extremely discounted rate. Great publicity for them, I naïvely thought. What an opportunity to do a good deed and score some major marketing points at the same time. It didn't take me long to realize that this wasn't going to happen. The call center folks at the Delta ticketing reservation center just weren't empowered to help me in any way—or at least that's what they told me. I inquired about bereavement fares but was told those are only discounted from the price of a last-minute ticket purchase, which is typically in the thousands of dollars. The price per ticket out of Ontario, California, into Great Falls was about $300 for an advance purchase. I decided to use some accumulated frequent flyer miles to get one of the tickets, and the other two I put on my credit card. There goes another $600, I thought.

It was at this point that I realized there was no sense in worrying about the cost of this whole project. Granted, I am not a wealthy person by any means, and in fact, despite making a decent salary, I do end up living paycheck to paycheck by the time all of my bills and expenses are paid for. Sure, I could use the money that I was spending on this project to buy a new shotgun or snowboard, but that's all it would be: something for me. At some points in their lives, people have to financially support the things that are important to them, whether that be a nonprofit organization, a political candidate or, in this case, helping to right a series of wrongs that occurred a long time ago. Now was the time. It's only money, right?

Friday, September 23, was on me before I knew it. Once our kids were out of school, we headed over the divide to Great Falls and rolled in to my parents' place at about 7:00 p.m. After getting everybody settled in, I drove to the Heritage Inn to check in with Dorothy and the rest of her group. Unlike with the airline, I did have some success getting a deal on lodging. The Volk family, who own the Heritage Inn, provided a room at completely no charge when I explained to them the situation about Herbert and his headstone ceremony. It was an act of kindness I have not forgotten.

I waited in the lobby for Dorothy to arrive. Unbeknownst to me, two of her other daughters had also made the trip, and I was introduced to them first: Pahmela and Carlotta. Soon thereafter, Dorothy rolled up in her wheelchair. I was immediately struck by how young she looked for somebody in her eighties. She was wearing a dark-blue navy hat that stated, "Korean War Veteran." With a somewhat mischievous gleam in her eye, Dorothy shared with me that on the plane trip up from California, a smallish, slightly hunched-over older gentleman had admonished her for wearing a hat she hadn't earned. She promptly told him that she had indeed fought in the Korean War and that he better be more careful about his presumptions. Apparently, the old man left humbled and better educated.

Dorothy and I chatted for about an hour, and then I departed. We made plans for her family to come over to my parents' house for dinner the following evening. It turned out to be an entertaining and engaging night. My wife's parents had also made the trip from Minnesota, and my brother and his family, who also live in Great Falls, attended as well, so there was plenty of conversation to go around. Sydney, Dorothy's granddaughter, entertained us with a selection of songs, which she sang a capella. She had been a recent contestant for the NBC show *The Voice*, and her talent was obvious.

On Sunday morning, my wife and I attended services at the Union Bethel AME Church with Dorothy and her family. The services were relatively informal and sparsely attended. The pastor of the church had recently departed for a position leading a congregation in Washington State, so it was left to some lay members of Union Bethel to direct the services. The leader of the youth group provided an update, noting that the youth group at that point consisted of only her three children but that "they were always praying for more members." Had it not been for my wife and I and Dorothy and her family, the number of attendees could have been counted on one hand. The services were quaint yet powerful.

The ceremony at Highland Cemetery was set for 2:00 p.m., which gave me and Christine time to grab a quick bite to eat after the church service ended. We did so, and then we headed to the graveyard. I wanted to give us plenty of time to find the headstone and to be there when other folks started arriving. I had been so busy with other things that I had forgotten to touch base with Bob Stewart, the head groundskeeper at Highland, to verify that the marker had been put in place. Just my luck, I thought to myself, all these people coming to the ceremony and the actual headstone might not even be installed. After a little bit of searching, we located Herbert's resting place. The marker had indeed been set. Breathing a sigh of relief, we hunkered under a tree in what little shade we could find. It was an aridly hot day, and not surprising for Great Falls, the wind was blowing stiffly.

Slowly, vehicles started to arrive. I noticed what looked like an older Dodge PowerWagon with what appeared to be a pumper unit on the back, but I figured it must be a cemetery vehicle of some sort. Only when it got closer did I realize it was indeed a fire truck, what is referred to as a slip-on Type 6 engine in the firefighter's vernacular. The writing on the door of the vehicle said, "Judith Basin Fire District." Inside were three men, all dressed in yellow Nomex fire shirts. The vehicle parked, and I walked over to introduce myself. The trio gave their names as Swede, John and Fred. They were from near Raynesford, a very small community located about thirty-five miles southeast of Great Falls. Swede told me they had read about the ceremony in the paper and had decided to come and pay their respects. Dorothy and her family were tremendously moved by their decision to attend, as was I. Firefighters at their best, honoring fallen comrades they never knew.

Others trickled in as well. Members of the congregation who had been at the church services earlier began to show, as did my parents. Bill Bronson, one of Great Falls' city commissioners, was there to read the official proclamation that had been drafted by the city and Mayor Robert Jones honoring the Forgotten Five. I had not requested such an action be taken. I was moved that they had chosen to do this and proud of my hometown. By official decree, September 25, 2011, was identified as "a day to remember the Forgotten Five and encourage all citizens to join in this salute to express our gratitude for the sacrifice these men made in service to their country." Chris Sorensen, from the National Smokejumper Association, attended to lend his support, as did Ken Robison, the historian who had been instrumental in helping me locate Dorothy. And of course, Dorothy was there with her family. Near the end of the events, Sydney, Dorothy's granddaughter, sang a poignant and moving rendition of "Amazing Grace."

When it fell to me to say a few words, I simply remarked that the day's ceremony was the culmination of the efforts of many people. I thanked them all for their contributions in helping to make the event possible. In closing, I reminded those assembled that there were two more individuals

The gravestone for Herbert Novotny. Highland Cemetery, September 2011. *Author's collection.*

lying in unmarked graves in Choteau and that the body of another one of the victims was still missing. We owe it to them, I noted, to bring a more fitting closure to their stories as well.

Postscript

A couple months after Herbert's ceremony, a large box arrived at our house. Inside, carefully packaged in bubble wrap and Styrofoam peanuts, was an elaborate, handmade dollhouse: a gift for my kids, Summer and Skye, from Dorothy. It was her way of saying thanks.

Later in the book-writing process, I realized I had made a mistake regarding Herbert's date of birth. His actual birthdate is January 28, 1911. The date on his gravestone is actually the date of birth of one of his siblings. When I realized my error, I became a little more understanding with regard to the inaccuracies on the marker in Boise at the Wildland Firefighter Monument.

When I was in Great Falls visiting my parents during the middle of July 2014, I was reading through the most recent edition of the *Great Falls Tribune*. It was here, in the obituary section, that I learned of the death of Lawrence "Swede" Malmberg. Swede had been one of the three firefighters who had come to Herbert's ceremony. He died at the age of seventy. I think of him often, even though that was the one and only time I ever met him. Like me, he had a passion for firefighting and owned an extremely large collection of dark blue fire T-shirts. Rest in peace, Swede. I will never forget your act of gratitude.

8

CHOTEAU CEMETERY

July 17, 2014

I have been driving around Great Falls with two dead guys in the back of my car for the better part of three days. I picked them up on Monday, and here we are on Thursday and they are both still there, taking up a large portion of the back of my beat-up Subaru wagon, not to mention the fact that their combined weight is causing my car's tired rear struts to sag like a wannabe rapper's jeans. I have a plan for getting rid of them, but the date for doing so is still nearly six weeks off. In the meantime, I have to figure out what to do with them until that time comes. As protection from inquisitive eyes, I have covered them up as best I can with the other detritus that has accumulated in the back of my car.

Since my son and I have been staying with my parents for the week while he attends a basketball camp here, I've contemplated stashing them somewhere in the garage, just to give my car's rear end a break. I fairly quickly discarded that notion. The things are heavy, which means I would need help, and that help would start asking questions and I just don't feel like answering any of those right now. I guess I'll just bring them back to Missoula and figure out a better plan as I go. I have a large garage, so chances are they will end up in some dusty corner, out of sight from prying eyes. Hopefully, my rear struts will hold out that long.

For the sake of transparency, it should be noted that I do not really have two dead bodies in the back of my car but rather *headstones* for two dead guys. Charles Allen and Hjalmer "Harry" Gunnarson are both buried in the Choteau Cemetery. Neither presently has a gravestone. The granite markers currently in my possession will address this deficiency. A memorial service at the Choteau Cemetery has been scheduled for August 24, one day before the eighty-third anniversary of their deaths in the Waldron Creek Fire.

August 24, 2014

The day breaks overcast and cold, with a light rain pushed by northerly winds—atypical weather for August. Fitting, though, for a funeral ceremony I suppose. It's hard to believe that this day has finally come. A decent amount of legwork has gone into trying to prepare for it. Years ago, I contacted the Teton County Courthouse, and it provided a detailed map of exactly where the men were buried (Block Number 6, Grave Number 6, Gunnarson in the north half, Allen in the south half), along with the rules and regulations for installing markers. Throughout the past year, I have been in contact with the cemetery's sexton, Jerry Kimmet, letting him know what I was doing.

Last fall, before the snow fell and the ground froze, Jerry poured the cement bases on which the markers would rest. The gravestones I ordered from Malisani, Inc. in Great Falls, the same place from which I obtained Herbert Novotny's marker three years earlier. Although I had paid for all of the costs associated with the Novotny ceremony (headstone and airfare for Dorothy, her daughter and granddaughter, et cetera), I chose to include some other folks when it came time to foot the bill for this affair in Choteau. It was a decision that took some effort. Asking for help is difficult for me, and I was hesitant to do so.

Part of the reason probably lies in the fact that I have some strong ownership in this endeavor. It's mine. I have been working on the Forgotten Five project for some time, and I did not want to have to rely on help from anyone else. Reaching out to others entails giving up some control, and I was not sure I wanted to do that. A couple of factors helped change my mind. One of those is the fact that I'm somewhat of a tight ass when it comes to money. It's not so much that I'm a penny pincher; rather, no matter how much money I make, I seem to be perpetually broke by the end of each month. Funding this entire Choteau project would only exacerbate that hard truth. The other element that helped sway me was the realization that there were several upstanding organizations and individuals who were willing to assist. Eventually, I reached out to some of them, and they stepped up to the plate with donations. The Board of Directors of the National Smokejumper Association, the National Museum of Forest Service History and Chris Sorensen all made rather sizeable contributions. I ponied up some of my own dough as well, which helped me deal with my control issues.

The service itself was fairly easy to schedule. I first called the Lutheran church in Choteau, since that is the denomination that my parents attempted to raise me in and thus the one with which I was most familiar. They were "between relationships" with regard to their pastor, since the previous one had taken a position elsewhere, and the new one would not be arriving until around the middle of August, roughly the same time I wanted to have this ceremony. A few people familiar with the area suggested I contact Ottis Bryan, a retired minister turned rancher from the local Choteau Baptist Church who lives just a few miles out of town. Due mostly to my own religious ignorance, the "Baptist" term freaked me out a little. My preconceived notions about Baptists incorporated "fire and brimstone" elements of the scripture, and I wasn't sure I wanted to go there. Pushing my prejudices aside as best I could, I called Ottis, and he graciously agreed to officiate. His selection would later reveal itself to be a wise one. He was the perfect fit: older, experienced, familiar with the local area and a top-notch storyteller.

The initial burial ceremony for the men had taken place on August 29 at the Roberts Funeral Home in Choteau, four days after their deaths in the fire. The September 3, 1931 edition of the *Acantha* reported them as "simple but impressive services." The article goes on to state that "Forestry officials and a small number of others were in attendance." It does not list who those people were. The Forest Service Supervisor's Office, located in Great Falls, was able to get in contact with Gunnarson's sister, who lived in Arborg, Manitoba. She asked about her brother's body being shipped to Canada but was told the government would pay the expense of shipping bodies of men killed fighting forest fires only to destinations in the United States. It would not pay the costs for shipping to foreign countries.

It still boggles my mind that such a policy could be in place and that someone from the government could even try to explain, in good conscience, that policy to a grieving family member. So apparently, the rules were such that anyone could be hired to fight the fire, but if he happened to be from another country and had the misfortune to die on said fire, he was shit out of luck for getting the government to cover the costs of sending his body home or paying funeral expenses. Amazing. Charles Allen, the other man killed in the fire, had given his mother's address in Pittsburgh, Pennsylvania, when he signed up to fight the Waldron Creek blaze, but she could not be located at the address he listed. Therefore, he was buried in Choteau as well. Truthfully, I wonder how hard they really looked for her.

I was curious to see how many people would attend the day's event. My wife and two children had accompanied me to Choteau, so I knew they would be present, as would my best friend, Jeff Hindoien; his wife, Jill; and their son, Luke, as well as Jeff's brother Chris and his wife, Tammy. Chris and Tammy live in Choteau, and both had been extremely helpful in various aspects of the project. Counting all of them, I knew at least nine people would be there. If that's all that came, it would be enough to be a respectable showing.

As we waited for the services to begin, various others started to arrive. The Forest Service was well represented, which meant a great deal to me. After all, these men had died while employed for it. Forest supervisor Bill Ivey was there, as was district ranger Mike Munoz and fire management officer Brad McBratney. In other words, some of the top leadership for the Forest Service, not just for the local district but also for the entire Lewis and Clark National Forest. I was impressed. Jim Phillips from the National Smokejumper Association was there with his wife, as was Larry Wright, an old smokejumping colleague of mine. He was retired from jumping but now ranches in the area. Dave Stack from the National Museum of Forest Service History was there as well, along with his wife.

The most striking figure of all was a rather tall gentleman, outfitted in the dress blues of a structural firefighter. I introduced myself to him and came to learn that his name was Russ Moorhouse. Russ was a retired chief from the South Placer Fire District in California. He and his wife, Nancy, had recently moved to Choteau, where Russ is now a member of the Choteau Volunteer Fire Department. When asked about why he attended this service, Russ simply said, "It was appropriate to do this. We needed to pay our respects." All in all, it was an impressive and diverse showing of folks who had come to pay their respects to firefighters Allen and Gunnarson. I was humbled that they chose to spend their valuable time by attending.

I had met with Reverend Bryan over dinner the previous night. He looked every part the Montana rancher, with boots, jeans and a black Stetson that remained perched on his head throughout the course of the evening. He shared with me his general theme for the service. It was quickly apparent that Reverend Bryan had spent considerable time preparing his remarks. He asked if I would share a few thoughts before turning the ceremony over to him, which I agreed to do.

On the day of the service, I kept my comments short since I wanted the focus to be on the two firefighters and on the words of Reverend Bryan. I thanked those present for their attendance and then borrowed a quote from author Nicholas Sparks, who once wrote in his book *Safe Haven*, "It's never too late to do the right thing." Nothing could be truer in this case. Gunnarson's body should have been sent to his sister in Canada. More effort should have been put into finding relatives of Charles Allen so that his body could have been shipped home as well. And lastly, headstones should have been provided by the government for these men when they were first buried. They were not. Now, nearly eighty-three years later, that was finally being done. It was not too late to do the right thing, and that is what we were accomplishing today. I quickly concluded my remarks and then turned things over to Reverend Bryan. Here are his words:

> *I'm now retired, but after fifty-plus years of active ministry, this is a first for me. I've been in many cemeteries in a few states and in Choteau for thirty-two years, but here that still qualifies me as a newcomer. During these years here, I have officiated several dozen occasions in this cemetery. But I have never conducted a service this long after someone's death.*
>
> *On most occasions, family and friends would gather around the casket over an open grave. The service would conclude, and the casket would be lowered into the open hole. At a later appropriate time, an engraved headstone would be placed at the head of the grave. On a few other occasions, family and friends would gather around a much smaller grave, perhaps a foot square, and a small box or urn would be lowered into the burial ground. Later, an engraved headstone would mark the spot.*
>
> *On Memorial Day each year, in this cemetery, the American Legion conducts a service for veterans who are interred here. The flag ceremony, the reading of the veterans' names who died the past year, the gun salute, the playing of taps are all a memorial and a tribute to the deserving men and their families. As a non-veteran, to be a part of that ceremony is a special moment.*
>
> *Often, after a funeral or memorial service, people would use the occasion to linger and wander through the cemetery to visit graves of loved ones. Sometimes they would spot a headstone with a familiar name and pause to reflect on the person's life and death. A story would be recalled about an event during their life. Many such stories would start out, "Do you remember when…?"*
>
> *Not here. No stories could be told over these two graves. Many have walked by this spot. Sometimes, even over it, but you can't fault them. There*

was no stone to remind them to be respectful and to walk carefully. Just sod—nothing sacred underneath the grass. No friends to come by to pause and reflect and say, "Do you remember when…?"

No one to recall there was that awful fire on August 25, 1931—such a tragedy. No one to say, "I remember reading about it when they found the bodies." None of that. No headstone and no stories told over this spot. It's just sod.

Charlie Russell would call this place a "trail plowed under." I'm guessing what he meant was that cows have trampled out the trails the buffalo had used since ancient times. Plows have destroyed much of the history of those who have gone before. Wooden houses have replaced hide teepees.

We see the truth of "trails plowed under" all around us. On our ranch, I heard of a family graveyard. I looked for it without finding it. One day, a ninety-year-old man showed up chauffeured by his nephew, and he wanted to see the place of his birth and the family cemetery. No house, and no evidence of the house; no barn and no evidence of a barn; no corrals, and no evidence of corrals on that spot of ground. It had been plowed for decades before I showed up to continue the plowing, planting and harvesting.

The family cemetery—that was a different story. He told me he knew exactly where it was, and sure enough, with his directions, we drove right to it. I never even knew it was there. He told me that he had infant siblings and cousins buried in it. His mother had built a little picket fence around the graveyard, maybe twelve-foot square, and had painted the fence white. She also had dug up a small lilac chute and planted it in the southeast corner. He was right. The little lilac chute was now a big bush that had spread out. Part of the picket fence was maybe two feet off the ground, hung up in the grown bush as it matured. "Trails plowed under."

We see evidence of this on the Old North Trail at the foot of the mountains west of Choteau. Only educated, knowing eyes can see a glimpse of the travois trail used for centuries. Now it is a path barely visible in places, plowed under by time and those who followed. The trail was recently marked, where visible, with large engraved stones. Schoolchildren of today make tours to see the past.

We're here today because of two trails that had almost vanished. We do have to ask how that happened. No family could be located. The family was too poor to have had the body shipped home. The government dropped the ball. These are the explanations offered.

Whatever the reason, a grave was adequate, but a headstone was deemed not necessary. Life moves on, driven by necessity and convenience. The sod takes over, and "trails are plowed under."

Necessity, convenience and neglect may have caused some human events to become lost, but they are never lost to God. Jesus spoke of His Father's attention to every detail of his creation. The Lord is no absentee landowner. He is a hands-on manager giving attention to the smallest details. To illustrate His Father's attention, Jesus chose the sparrow. The sparrow is as tiny and insignificant as the eagle is majestic. The sparrow is as small as the ostrich is large. The feathers of the sparrow have not one hint of the peacock's plumage. The sparrow has no song to sing like the mockingbird. He is what God has made. For that reason, when the sparrow falls to his death, God stops to take notice. If God so cares for the sparrow, what about the creatures created in His image? Those He crowned with glory and honor. Those He placed in charge of His creation. Those He created for eternity. Those He calls sons and daughters.

For His second illustration of His Father's attention to His creation, Jesus used the lilies. No temple, no king's palace, no skyscraper, no Taj Mahal can match the creation of the lilies. Zillions are scattered around the world and God is the gardener for each one. Yet His lilies bloom for a day or two then go the way of all grass, never leaving a sign.

Since the creator watches over His sparrows and tends His lilies, what must we think? What do you think? What do you think His attention is to those He created in His image to be His sons and daughters for all eternity?

On August 25, 1931, five men died in the Waldron Creek fire. I assume the fire and the deaths went unnoticed by most in the United States. Jesus said not so with His Father. God sees every sparrow that falls and He is the gardener of every lily on earth. "For God so loved the world that He gave His one and only son, that whoever believes in Him shall not perish but have everlasting life." (John 3:16 NIV)

He saw the five die as He saw the Unknown Soldier die in some nameless battle on an unfamiliar battlefield and is now buried in Arlington National Cemetery. Surely some of you have been there for the changing of the guards. Every day, twenty-four hours a day, the Unknown Soldier's comrades march. It has become not a place of death but a place that contributes to the spirit of this nation, to a healthy national consciousness. We need to celebrate the Unknown Soldier to remind ourselves what is most noble about the human spirit.

I believe a little of the spirit of the Unknown Soldier goes out from occasions like this. We have a need to honor the dead. We have a need to celebrate the best in the human spirit. Today's firefighters need to know of their comrades across the years that died, literally, in the line of fire.

We stand before two trails almost plowed under. Dr. Palmer snatched up barely visible trails. He kept scratching and following leads, and they led over several states, through old newspapers and into courthouses' dusty records. The trail led here to where we now stand. These graves and names of two men have been brought together with headstones, and that is a long chapter but not the final chapter. People who wander through this cemetery will say, "Oh yes, I remember reading about that terrible fire. It happened just west of Choteau. Two of the five are buried here. If you've never heard of it you need to read the story. Look! Here are their names." Trails plowed under, almost. Almost.

I was surprised but not shocked when I learned that Reverend Bryan was utilizing Charlie Russell's "Trails Plowed Under" theme for his service. He had told me when we met for dinner that this was the premise on which his graveside service would be based. I was excited to hear how he planned to incorporate it into his sermon. In 1927, a year after Russell's death, a book with the same title was published. In it was a collection of short stories written by Charlie himself on a variety of topics, ranging from wolves and wild horses to bad men and buffalo hunts. In the introduction to "Trails Plowed Under," which Russell penned just a few months before his death, he wrote:

Life has never been too serious with me—I lived to play and I'm playing yet. Laughs and good judgment have saved me many a black eye, but I don't laugh at other's tears. I was a wild young man but age has made me gentle. I drank, but never alone, and when I drank it was no secret. I am still friendly with drinking men. My friends are mixed—preachers, priests, and sinners. I belong to no church, but am friendly toward and respect all of them. I have always liked horses and since I was eight years old, have always owned a few. I am old-fashioned and peculiar in my dress. I am eccentric (that is a polite way of saying you're crazy). I believe in luck and have had lots of it.

I never get too far away from Charlie Russell. He is always there for me, it seems. We share the same first name, of course, but our relationship goes much deeper than this.

Charles M. Russell sitting with a painting. Date unknown. *University of Montana Archives and Special Collections.*

As an elementary school student, we used to take field trips to the C.M. Russell Museum in Great Falls to see his artworks and learn about his life. When scattered cousins, aunts and uncles from across the United States would visit my family during their summer vacations, our sightseeing tours usually included a stop at the museum as well. As I began to work on this book, Russell continued to make his presence felt. At Highland Cemetery, he was there as the efforts to put in a headstone for Herbert Novotny took place. Since he was buried in Highland in 1926, five years before the men in Waldron perished, I would like to think that Charlie was watching over them as they were laid to rest.

When I made research trips to the Montana Historical Society in Helena to look through its microfiche copies of old newspapers, he was there as well. The Research Center, which houses the materials I needed, lies on the second floor of the complex. On the first floor is the C.M. Russell Gallery, a two-thousand-square-foot exhibit composed of nearly eighty Russell art pieces, including oils, watercolors, pen-and-inks, pencil sketches, sculptures, bronzes and even illustrated letters that he wrote to some very fortunate recipients. After hours of sitting and reading old newspaper articles, I would take a break, head downstairs and wander around the gallery, amazed by

The gravestones of Charles Allen and Hjalmer Gunnarson. Choteau Cemetery, Choteau, Montana, August 2014. *Author's collection.*

Russell's ability to capture what he had seen. It felt good to be this close to him. I imagined he was watching me, keeping an eye on my progress with this book.

Therefore, it seemed perfectly fitting that Reverend Bryan based his graveside service on a concept from Charlie Russell. It was almost as if fate, or something very much like it, had played a hand. When the ceremony concluded, the loops in Choteau, at least with regard to headstones, had finally been closed. Trails that had been plowed under were now excavated, once again visible for the whole world to see.

9

ON THE FIRE GROUND

Even after several years of work on this project, I had never actually set foot on the site where the fire burned and the five men died. I had, though, been near it on numerous occasions and for a variety of different reasons. My best friend, Jeff Hindoien, had grandparents who lived in Choteau, so during our high school and college years we spent a considerable amount of time in the approximate area of the fire, hunting, fishing and skiing at Teton Pass Ski Area. Despite being so close to the place, I had no idea at the time that five firefighters had been killed in a fire at Waldron Creek. To this day, there is no interpretive signage about the fire alongside the Teton Canyon Road that runs near where the men perished.

On a Wednesday near the end of November 2014, Jeff and I made plans to hike up into the area where the fatalities occurred. He took the day off work because that's the kind of thing best friends do for each other, even if the plan is to trek some of the nastiest, steepest terrain on the east side of the continental divide. Jeff is a lawyer by trade, and a damn good one at that. Although he has no experience as a wildland firefighter, Jeff is an outdoorsman, and he combines keen observational skills with a sharp, analytical mind. When it came to finding a partner for the excursion, he was the perfect choice.

Although the blaze has always been referred to as the Waldron Creek Fire, in actuality most of the fire was confined to the drainage that forms the South Fork of Waldron Creek. The North Fork of Waldron Creek, located in the next drainage to the north, and the South Fork merge farther down

Forest Service signage of The South Fork of Waldron Creek along the Teton Canyon Road. *Author's collection.*

Map courtesy of Google Earth.

canyon to form Waldron Creek, which eventually flows into the North Fork of the Teton River. Jeff and I drove the Teton Canyon Road and parked the truck at the gate of Forest Service Trail 193. This trail runs just to the north of and parallel to the South Fork of Waldron Creek. We hiked in, and after about eight-tenths of a mile, we left the path, jumped across the skinny South Fork and headed cross-country toward the top of the drainage, a little more than two miles away and 1,600 feet higher in elevation. At the time of the fire in 1931, the Teton Canyon Road did not exist, which meant that those getting to the fire had to park at Deep Canyon Ranch, near the mouth of the canyon, and walk the approximately twelve miles to the fire, the last two of which would leave even a mountain goat feeling slightly asthmatic. Thankfully, we had the luxury of driving the road.

Since it had snowed a couple of days earlier, footing was a challenge, especially in the steeper parts of draw. Both of us carried rifles and wore the requisite hunter orange clothing in the off chance that we might encounter an elk. Although we saw a few old deer tracks in the creek bottom, as we climbed higher and higher, all signs of game ceased. Even if we had seen something, the thought of having to pack out a dead animal over this terrain was sobering. We vowed that it would have to be a trophy bull if bullets were going to start flying.

The process of writing this book was very abstract at times. It was tough to believe the deaths even happened, and it often felt like I was working on a piece of fiction. Since I had never walked the fire ground where the men died, the whole affair often seemed more theoretical than concrete. I assumed that when I actually hiked around Waldron Creek this feeling would change. And it did, but at the same time, it did not. The standing and fallen dead snags that are still there helped me visualize how and where the fire burned. The steep slopes and the difficult footing made it easier to understand how challenging it must have been for the men to navigate the landscape. However, with no crosses to mark where the men fell, their deaths still seemed intangible. I had no idea where exactly they died, and with this vital piece of information missing, it contributed to the abstractness.

Since I am fascinated with fatality fires, I have had the opportunity to explore places like Mann Gulch in Montana and South Canyon in Colorado. Together, these two blazes claimed the lives of twenty-seven firefighters. At the site of both of these incidents, markers on the fire ground clearly identify where each firefighter died, and an informal network of trails makes it somewhat easier to move around the precipitous landscape. In fact, in 1997, as a Missoula smokejumper, I had the opportunity to help install new granite

Getting ready to place the new marker for Robert J. Bennett. Mann Gulch, Montana, June 1997. *Author's collection.*

monuments in Mann Gulch for each of the thirteen men who perished. The old white crosses that had been installed shortly after the fire in 1949 were in varying states of decay. While some were still in relatively good condition, others had gotten to the point where the crumbling concrete had exposed the rebar skeletons inside. After considerable debate, we decided to place the new markers right next to the old ones, leaving the old ones in place to continue losing their battle with the elements. More than anything else, the monuments serve as physical reminders of death. You can see just how close, or how far, each firefighter was from the top, from safety, from life.

Smokejumpers Mike Patten (left) and Jim Beck (right) place the marker for Joseph Sylvia. Mann Gulch, Montana, June 1997. *Author's collection.*

In Waldron Creek, with nothing in place to designate where the men fell, it was just harder to believe that people were actually killed here. When Jeff and I wandered around the gulch, I had to keep reminding myself that five men really did die in this vicinity. And although I have many years of experience as a wildland firefighter, I am far from being an expert in fire behavior, so it was not like I could look at the terrain and figure out exactly where the men were when they fell. Complicating matters is the fact that there was a previous fire in the same area in 1917. This makes it even more difficult to determine which dead snags and old fire scars belong to which of the two fires.

The newspaper accounts after the tragic event repeatedly asserted that the spot fire where the twenty recruits were left was about four hundred yards to the east of the main fire, on the opposite side of the gulch and hidden by a small ridge. Coroner Roberts emphasized that the bodies were found at 7,500 feet of elevation. Therefore, if Roberts was accurate in his estimation of the elevation, it puts the fatality site very near the top of the drainage. However, the piece of the puzzle that does not make sense is the account about the spot fire supposedly being four hundred yards east of the main fire and on the opposite side of the gulch.

South Fork of the Waldron Creek drainage, where the fatalities occurred. *Photo courtesy of Jeff Hindoien.*

On the actual ground, these figures do not fit. At the 7,500-foot elevation mark, if the spot fire was on the east side of the gulch and the main fire was on the opposite (west) side, that distance is only about two hundred yards. The accounts also noted that the main fire was hidden by a small ridge. Based on our on-site observations, there is no small ridge within the upper portion of the drainage. The only ridge is the much more predominant one that serves as the western boundary of the entire gulch. Therefore, several different scenarios could be possible: the spot fire and main fire were on opposite sides of the same gulch and the distance figure between them of four hundred yards quoted by the newspaper accounts was just grossly inaccurate; the main fire was actually in the next drainage to the west; or the main fire and the spot fire were in the same drainage but much lower than the reported 7,500-foot elevation. At this point, I don't know enough to even guess about which of these is correct.

With daylight waning quickly, Jeff and I decided that we'd better start heading down and back to the truck before it got too dark to see. We each snapped a few pictures of the drainage, and then we began a somewhat treacherous descent over ice-covered rocks. Although it was an insightful outing, many questions were left unresolved. Another trip into Waldron Creek was going to be necessary, that much was clear.

Round Two

The second trip into the site of the fire was planned for May 2015. I had wanted most, if not all, of the snow to be gone so that navigating around the fire ground would be less difficult and the landscape unobscured. The previous winter had been relatively mild, with a snowpack that was well below average, but late spring storms dumped fair amounts of the white stuff in the high country, scuttling those May plans. By the time my schedule allowed a return visit, it was early July before I was able to get back into Waldron Creek.

Due to work demands, Jeff was unable to accompany me on this second trip. Although I was bummed that he was not able to join me, the more I thought about it, the more fitting it seemed that my return to the site would be a solo venture. So much of this project I had completed alone. Therefore, I thought it seemed apt that my final visit would leave me there by myself. Knowing it would be a long day, I left Missoula early in the morning. The month of June had been unbearably hot, breaking historical records for heat and lack of moisture, so I was surprised when I ran into rain around Ovando and even more shocked that it continued to fall steadily for the remainder of the drive.

As I drove up the Teton Canyon, I tried to put myself in the shoes of the Forgotten Five, what they must have been feeling and thinking as they hiked their way into the fire that would eventually claim their lives. Based on my own experiences, it must have been some combination of apprehension and excitement, for they were heading into a grand unknown. The rain continued to fall as I made my way up canyon. The land needed it, and whatever amount fell would help decrease the chances of new fires, at least for a little while.

It was pouring by the time I parked at the trailhead. I traveled light, carrying only some water, bear spray, a digital camera and five red roses that I had purchased before leaving town. Within minutes of the start of the hike, I was soaked to the skin. In my initial plan, formulated immediately following this first visit to Waldron Creek, I had planned on hiking all the way to the top, with the goal of doing my best to determine, if I could, the approximate spot where the spot fire burned and the men fell. But in the ensuing months since that first visit, and after careful consideration factoring in many details, I ditched that idea. I realized that it was not necessary to figure out the position of the spot fire or where the men died, nor was either of these even feasible. Too much information

A rock cairn and five roses at the bottom of the fatality site drainage. *Author's collection.*

was missing to identify these locations. And most importantly, their story is better served by having the whereabouts of both remain a perpetual mystery. It highlights just how overlooked and forgotten their sacrifice has been and how poorly it was investigated after it occurred.

At the bottom of drainage where the men died, at about the spot I presumed would be the most logical place for them to have begun their final fateful climb up to the spot fire, I constructed a small rock cairn. I took my time, trying to find the best stones to make it as tight fitting and sturdy as possible. When finished, I leaned the quintet of flowers across its face, closed my eyes and began to mumble an awkward prayer for the Forgotten Five.

In that moment of solitude, with eyes shut, I envision them at that critical moment in time just as the Waldron Creek Fire rears back to make its deadly run. The wind freshens. Thick, black smoke rolls up the drainage, and brilliant orange flames follow closely on its heels. "Follow me. We can get this thing!" someone shouts. Because of the now-reduced visibility, I cannot tell who orders this call to action, but in response to it, the Forgotten Five separate themselves from the pack. They do not panic and attempt to flee the mushrooming conflagration, as some of the survivors will later attest. They decide to fight.

The men grab their tools and start heading toward the burgeoning blaze below them. They cast a final, contempt-filled look at the fifteen who have remained behind. Standing there among them, a prescient smirk on his face, the dark stranger watches the five intently as they leave. Hjalmer Gunnarson, the Canadian veteran of the Great War, appears to be in charge. The men respect Gunnarson, for they know from talking with him during the day that he has grown up fighting fires in his native Manitoba, and his experiences in the trench warfare of World War I have made him cool under fire.

Like a mother clutching her newborn child, the main fire and the spot fire pull together into one. Flame lengths double, then quadruple, in size. Fire whirls, small tornadoes of combustion, catapult the inferno skyward. Burning embers rain down, and when they land, new blazes spring to life in a mesmerizing, albeit lethal, process of reproduction. The Forgotten Five begin their attack, scraping the earth down to mineral soil, wanting to starve the ravenous beast of the fuel it needs to survive. Fire closes in around them.

They work in unison, but the meager fireline they are able to scratch out is no match for the flames, which roll across the terrain like a crimson wave. Sweat pours from their brows. The men are slowing now as the super-heated gasses and lack of oxygen begin to take their toll. They huddle together in a small trench, a near final act of solidarity against the impinging firestorm.

Then, through the fumes, I see a solitary figure break from the others, and he begins a desperate sprint up the slope. For a brief second, there is a partial break in the smoke, and I see that it is Novotny who is attempting to flee. I am amazed at how fast he is able to move, especially considering the steepness of the climb and the conditions at hand. He makes it nearly the length of a football field, weaving his way through the maelstrom, before a swell of heat and flames tackle him to the ground. Mercifully, smoke rolls in, obscuring my view of the deadly conclusion. Gunnarson, Bierchen, Allen and Williamson are already dead, their airways singed taut.

From the top of the ridge, the dark stranger watches this final scene unfold. After its tragic completion, he disappears over the peak and rejoins the other fourteen in their retreat down the Middle Fork of the Teton drainage and out of the fire zone. Each of the men will have his final moment on this earth, and the dark stranger will be there to witness these, as well, but that time is not today. Not today.

10
THE SEARCH FOR TED

The Bierchen family embodied the American dream, their story a microcosm of immigration to the United States around the turn of the twentieth century. In all, five Bierchen brothers left their tiny village in Eschdorf, Luxembourg, and headed to Illinois in search of opportunity and a fresh start. Nicolas "Nick" Bierchen was the first to come, sometime around 1900. He and his wife, Lena Ebert, took over her family's restaurant on Ridge Avenue in the recently annexed and still-rural West Ridge neighborhood of Chicago. The restaurant also had a bar, a bowling alley and an outside picnic area. Nick and Lena had four children, all of them born in Chicago. Nick was active in the Luxemburger Bruderbund, a fraternal organization that focused primarily on social needs and maintaining the ethnic traditions of its Luxembourg homeland. Nick died of a heart attack during a card game in 1916 at the age of thirty-five. Lena and her brother Joseph continued to operate the restaurant until Joseph died in a Chicago heat wave in July 1934. After Joseph's death, Lena was sole proprietor until she sold the restaurant in 1939, whereupon it was named Allgauer's.

Jean, or "John," Bierchen, the next brother to come, arrived in the United States in early 1902 and immediately went to see Nick at his home in the Chicago area. John eventually made his way to Yankton, South Dakota, and he appeared in the 1905 state census there. He returned to Chicago; married Marie Schuller, also from Luxembourg, in 1910; and eventually owned and operated several greenhouses just north of Devon Avenue in Chicago. Marie and John had seven children, all of them born in Chicago as well. John died

John Bierchen. Date unknown. *Photo courtesy of Jim Heckenbach.*

suddenly of a stroke at the age of fifty on August 25, 1932, exactly one year after the death of his brother Ted on the Waldron Creek Fire. John is buried at Memorial Gardens in Skokie, Illinois.

Jacques, or "Jake," Bierchen and his girlfriend, Elizabeth Schoos, came to America in 1906. Jake's 1917 draft card noted that he was working as a bartender and assistant manager at Ebert's, the same restaurant that his brother Nick was operating. Jake, too, was active in the Luxemburger

The graves of Jake Bierchen and his wife, Elizabeth. St. Henry's Cemetery, Chicago, Illinois, December 2014. *Author's collection.*

Bruderbund in the Chicago area, helping to organize the *Schobermesse* festival, a celebration of the harvest season each Labor Day. Jake and Elizabeth were married in 1919. The following year, they were living on North Clark Street in Chicago, and Jake was working in a factory. He later was employed as an iceman. The introduction of electricity and refrigerators spelled the end of this career, forcing Jake into a diverse array of other jobs throughout the rest of his life to make ends meet. Jake and Elizabeth lived the rest of their lives on Ravenswood Street in Chicago and were active members of the St. Henry's Catholic Church. Collectively, Jake and Elizabeth had five children, and like their Bierchen cousins, all were born in Chicago. Jake died in 1960, and he is buried in St. Henry's Cemetery in Chicago. Elizabeth died in 1982, and she is interred by her husband's side.

Jean Pierre, or "Jim," was the youngest of the Bierchen brothers to come to America. He, too, was employed for a period of time at his brother John's greenhouses in Chicago. His World War I draft card noted that he was working as a farm laborer in Terril, Iowa. By early 1917, reports had him owning a farm in North Dakota, but he later leased it out and took a trip to California. From January 1918 until April 1919, Jim served as a private in the U.S. Army with the 82nd "All American" Division.[32] He completed two six-month tours of duty in France, fighting Germans during World War I. After the war, Jim moved to the Portland, Oregon area. He never married, died in 1971 and is buried in the Willamette National Cemetery.

Theodore, or "Ted," Bierchen immigrated to the United States in 1902, at the age of nineteen. At the time of the 1910 U.S. census, he was living with his brother John and working at his greenhouses in Chicago. Like his brother Jim, he made his way out to California for a period of time. Ted's World War I draft card listed him as living in Casselton, North Dakota. As noted previously, it is unknown if Ted was ever called up to active duty to fight in the Great War. The 1930 census showed Ted living in Cascade, Montana, and staying at a boardinghouse there while he worked at a local dairy. In Luxembourg, Ted's father, Nicolas Bierchen, had been a *feldhüter*, or guardian of the fields. Perhaps this familial link played some role in his decision to volunteer for the Waldron Creek Fire. Whatever the reason, it would prove to be a fatal choice.

I learned all of this information about the five Bierchen brothers from the research efforts of Jim Heckenbach. In 2003, when I first started looking into the Waldron Creek Fire, I had done a simple Google search for the "Bierchen" last name since it is so unique. One of the earliest hits was for the "Bierchen Genealogy Page," which turned out to be administered by Jim. I e-mailed him, telling him about the Waldron fire and that one of those who died was Ted Bierchen. Jim replied very quickly, informing me that Ted was his great-uncle. Jim's mother, Mary, is the daughter of John Bierchen, Ted's older brother. I use the present tense, since Mary, at age ninety-three, is still very much alive and living in Florida at the time of this writing. According to Jim, Mary has some distinct memories of Ted but has no recollection of ever attending a funeral for him.

The wedding of Jake Bierchen, October 1919. Ted Bierchen is third from left. *Photo courtesy of Jim Heckenbach.*

Articles in the *Choteau Acantha* and the *Great Falls Tribune* following the fire reported that Ted's body was shipped back to Chicago, accepted by his brother Jake and buried in St. Henry's Catholic Cemetery on the corner of Devon and Ridge Avenues in the northern part of the city. St. Henry Church, a neo-Gothic structure, was erected in 1905. It originally served the German Catholic community of the Windy City and is now the only surviving churchyard, which means a graveyard adjacent to and affiliated with a church, within the city limits of Chicago. Local lore has it that "Sailor Jack," the boy on the Cracker Jack box, is buried there.

Unfortunately, Ted isn't where he is supposed to be, at least not according to records. St. Henry's, which was given to a Croatian parish in the 1970s and is now called Croatian Mission of Blessed Alojzije Stepinac, has no documentation of Ted being buried there. Nor does any other Catholic cemetery in the Greater Chicago area. With the advent of the computer, individual parish burial records were conglomerated into one large database administered by the Archdiocese of Chicago. In none of these archives is there a registered burial for Ted Bierchen. I searched websites and databases,

I called St. Henry's and the Archdiocese of Chicago sacramental records departments, but the answer was consistently the same: we have no record for the burial of a Theodore "Ted" Bierchen.

I had to go. I owed Ted that much. Despite the fact that I had consistently been told there was no record of him being buried in St. Henry's, or anywhere else for that matter, I needed to visit Chicago and check for myself. Truthfully, I had little confidence that I would find anything useful. It was more a matter of the fact that the other members of the Forgotten Five had gotten the consideration they deserved, so it would have been unfair to Ted to exclude him from similar attention. Over Christmas break in December 2014, our family was going to be spending the holiday in Minnesota, so I decided that would be a good opportunity to fly to Chicago for a quick overnight trip. I e-mailed William Shatner's folks at Priceline.com, and they hooked me up with a cheap round-trip flight out of Minneapolis departing on the twenty-ninth, a hotel room in a Howard Johnson's by the airport that ended up smelling of cigarette smoke and sex and a rental car. Thanks, Bill, I owe you one.

I let Jim know I was coming, and he graciously agreed to take the train down to the city from his home in Lindenhurst, Illinois, and accompany me on my search. Although we had never met in person before, we quickly found each other in the baggage claim area and headed over to the rental car center. We agreed that St. Henry's was the most logical place to begin our search. Jim knew the city well, so I drove while he navigated us around on side streets, avoiding the congestion—and the tolls—of the Interstate.

The cemetery itself is small, about two and a half acres in size, and roughly rectangular in shape. Luckily for us, the weather that winter had so far been quite mild, meaning there was no snow on the ground, which would have obscured many of the gravestones. Jim and I started on opposite sides of the place and worked toward the middle, each of us looking for Bierchen markers. Checking stone by stone, I came across groups of other Luxembourg families who had also immigrated to the States in search of a better life: the Langs, the Schweigs and the Yotts, to name but a few, but no Bierchens. Jim was having more luck on his side, finding six Bierchen grave sites in all. However, none of them was the grave site for Ted.

I made a discovery at the far back of the cemetery, along the fence line that formed the western boundary of the property. Her flush-to-the-ground gravestone was almost completely obscured with dead leaves that had fallen and accumulated from the gnarled old oak trees scattered nearby. Catherine Bierchen was her name. Born on May 11, 1918. Died on June 30, 1919.

The gravestone for Catherine Bierchen. St. Henry's Cemetery, Chicago, Illinois, December 2014. *Author's collection.*

Barely one year old, she was the infant daughter of John and Marie Bierchen, thus making Catherine a sister of Jim's mom, as well as Ted's niece.

According to the records of St. Henry's Cemetery, while several of the Bierchen clan are listed as buried there, Catherine is not among them. Her name does not appear in the database. And yet here she was, her tiny gravestone caked with mud and leaves but otherwise intact. Family rumor had it that she had been dropped by an aunt who had come to help care for her, and this led to her untimely death. Little else is known of her short time on the planet.

Since Catherine is not listed in the records as being buried in St. Henry's, Jim and I decided to drive over to Calvary Catholic Cemetery in nearby Evanston. It is here that the centralized records of all Catholic burials in the

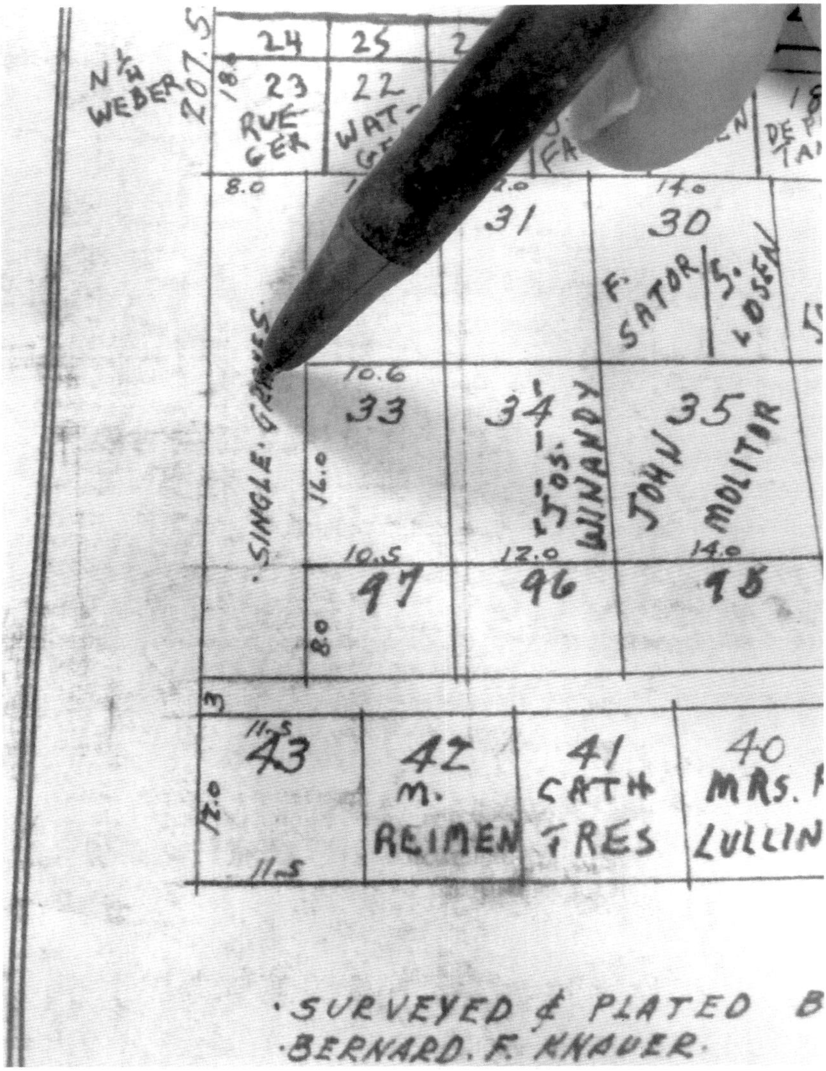

The single graves section of St. Henry's Cemetery. The pen points to the approximate location of Catherine Bierchen's grave site. *Author's collection.*

Greater Chicago area are kept. We explained the situation to the sexton on duty. After we showed him a picture of Catherine's gravestone and explained where in St. Henry's it was located, he began digging through some old file cabinets. A few moments later, he produced an old blueprint of the entire cemetery grounds.

The sexton explained that Catherine's marker is located in the "single graves" section of the cemetery, a portion of St. Henry's that had been reserved for charity burials. When I asked him to explain what this meant, he noted that the lots in this small part of the cemetery had been set aside by the church for the burial of those who lacked the financial resources to be able to afford funeral expenses. It would make perfect sense that Ted would be buried in this section, perhaps right next to his infant niece. The next closest grave marker to baby Catherine on a north–south axis is over twenty feet away along this western edge of the cemetery that was reserved for these single graves.

According to the cemetery's own records, this part of the cemetery is full, meaning there are plenty of unmarked graves near Catherine's. If the church's records are incomplete enough to have not included Catherine's interment, then it makes it much more likely in my eyes that it failed to record Ted's burial as well. Jake, Nick and John Bierchen and their families were all parishioners at St. Henry's Church. Newspaper accounts identified Jake as the person who received Ted's body when it was shipped to Chicago. And those same newspapers noted that Ted was buried in St. Henry's. The signs all seem to point to the conclusion that Ted is indeed buried here, even though this is not reflected in the church's records. Jake was making barely enough money to keep his family afloat. Since he probably did not have the resources to buy a headstone, and obviously the Forest Service had no interest in buying one for him since it had failed to do so for any of the other victims of the Waldron blaze, Ted probably became the fourth member of the Forgotten Five to be buried in an unmarked grave.

It would have been great to find a grave site for him, to bring a little bit more closure to this whole process. If not that, then at least a record of burial that identified the specific spot in the cemetery where his body lies. Then, like I did with the others, arrangements could have been made to get a headstone in place for him. Alas, neither of these is the case. However, I am confident in my conclusion that he is buried at St. Henry's. He will know that I came, that I looked for him, and that is enough for now.

11

MOP UP

The full and complete narrative of what actually happened on August 25, 1931, at the Waldron Creek Fire will never be known. Poor investigation and record keeping after the incident, coupled with the passage of time, conspire against obtaining a comprehensive picture of that day's tragic events. There are simply too many questions and too few answers. What remains are, at best, educated guesses, and even some of those may not be very intelligent. Sadly, the men who could have really helped us understand what transpired on the mountain that day cannot do so, for they were burned to death in the blaze.

The goals of this project were relatively simple: to introduce readers to the Waldron Creek Fire and the five forgotten individuals who died fighting it and to achieve a better sense of closure, if that is even possible more than eighty years later, for these five men. To accomplish these aims, headstones were placed for three of the men whose graves did not have markers. A memorial stone for all five was set at the Wildland Firefighter Monument in Boise. And the five deaths are finally registered in the Historical Wildland Firefighter Fatalities database.

But it has not been a perfect effort. One grave site—Ted Bierchen's—remains unfound, although I am fairly certain he is indeed buried where the records indicate. Unfortunately, even if he is in St. Henry's Cemetery, short of exhuming multiple graves and conducting DNA analyses on them (which I don't see happening), it will never be known exactly where he lies. The marker at the Wildland Firefighter Monument in Boise contains errors, the most notable being that the year listed for the fire is wrong. And no memorial is yet in place near where the men fell. It is my hope that some of these deficiencies can be addressed in the future, realizing that particular ones will be much more difficult to accomplish than others.

The Waldron Creek Fire occurred at or very near the historical nexus of wildfire suppression developments in the United States. Key advancements were happening, or were soon to occur, and these changes would radically alter the firefighting landscape. While airplanes had seen very limited usage for fire detection even as early as 1917, their capabilities had yet to be fully explored. By 1931, visionaries in aviation were beginning to conceptualize the many different ways airplanes could be used to aid in fire suppression, and they had begun to put many of these ideas into practice. This included using planes not only for fire detection but also for the rapid movement of firefighters and supplies to places of need.

When Waldron Creek burned, radios were starting to become a valuable tool for communication in the woods and on the fireline. Prior to this, the Forest Service had experimented with everything from carrier pigeons to signal mirrors for sending messages. Each had significant and predictable drawbacks. By 1931, usage of the telephone had become ubiquitous within the agency. It was a handy device for sure, but one dependent on extensive hardware and miles upon miles of wire to tie the system together. Fires did not seem too concerned about burning only where there was telephone reception. And the outbreak of incendiarism during the fire season of 1931 also brought with it the frequent practice of cutting telephone lines to hamper communications, increasing the chances that fires would grow larger and need more men to fight them. The radio solved these problems and many others.

The most important change that came on the heels of Waldron Creek was the development of "standing armies" of firefighters by the U.S. government. Within two years after the fire, the Civilian Conservation Corps had thousands of men available for fire duties. By the end of the '30s, in addition to these CCC firefighters, the Forest Service had initiated the smokejumper program and, by the following year, 1940, would have available a small number of smokejumpers from bases in Montana and Washington and the beginnings of what would eventually become today's hotshot fire crews. The loss of the five men in Waldron Creek helped, I believe, to spur this transformation.

The deaths of the Forgotten Five highlighted the fact that the early Forest Service's track record of taking care of those people who were hurt or killed trying to put out its fires was far from exemplary. This institutional lack of empathy and support was on full display after the fires of 1910 and continued to be exhibited following the deaths in the Waldron Creek Fire. To not pay for headstones for the Forgotten Five, or to even pick up the costs of shipping all the bodies home to their families, is, in my opinion, unconscionable. In its defense, the Forest Service has improved considerably in this area today. But progress has taken a painstakingly long time. And it has come at the expense of dead firefighters and their grieving families.

It is my conclusion that the Waldron Creek Fire fatalities, in addition to the fifteen other line-of-duty deaths during the 1931 fire season, laid the groundwork for significant future changes within the Forest Service. They were, in a sense, the earliest proverbial straws placed on the camel's back. The deaths of twenty-nine firefighters in 1933 on the Griffith Park Fire in California, followed by the loss of fifteen CCC and Forest Service firefighters in 1937 on the Blackwater Fire in Wyoming, were the stalks that finally broke its spine and at long last brought the dangers of firefighting into the public's conscience. The Forest Service was now compelled to more fully address the topic of firefighter safety. This included analyzing not only how firefighters were hired but also how they were trained, led and equipped. All of these early improvements, in my mind, are due in part to the contributions of the Forgotten Five.

In the first chapter of this book, I commented that the five men who perished in the Waldron Creek blaze died without knowing that their efforts made an impact. This much is accurate. When it comes to the influence they have actually had on others, though, it has been deep and far reaching. The two young daughters Herbert Novotny left behind would both go on to serve their country honorably in the military and to make significant other contributions to society as a whole over the course of the their lives. Dorothy, the younger of the two, continues to do so even now, well into her eighties. It is unknown if any of the other four men had children. Chances are at least one of them did. Who knows what influence on humanity these unknown offspring might have had over the course of their lives.

The legacy of the Forgotten Five also lives on in today's wildland firefighters. Now, thousands of them are hired each year not only to protect us from wildfire but also to serve as one of this country's first lines of defense against and recovery from all sorts of different incidents and disasters. Thanks in part to the ultimate sacrifice paid by the Waldron quintet, these individuals receive better training, leadership, equipment and compensation. And should any of them be injured or killed on the job, the entities that hired them, in addition to organizations like the Wildland Firefighter Foundation, do a significantly better job of supporting them and their families. This did not happen for the Forgotten Five.

Lastly, the Forgotten Five have left a tremendous impression on me as well. As a firefighter myself for nearly twenty years, I, too, benefited from their sacrifice. The fact that my position even existed was due in part to them.

Old Sawmill District. Missoula, Montana, February 2015. *Author's collection.*

At a much deeper level, through their deaths, I have had the opportunity to examine my own life. It is difficult to immerse yourself in the deaths of others without at some point contemplating your own mortality. Slowly, painstakingly, I am coming to grips with my own impermanence. Death is one of life's few guarantees, a constant we cannot change. Through the process of writing this book, I have learned that in order to fully appreciate life, I must be more cognizant of the fact that death awaits us all, me included. More and more, I am able to savor each moment as it comes, for there may never be another. For this gift from the Forgotten Five, I am eternally grateful.

As noted at the beginning of this book, I was peculiarly drawn to the Old Sawmill District and Silver Park in particular. I visited here frequently during the last two years of my writing this manuscript, but for most of that time I never really understood the reasons behind my actions. Eventually, I gained better insight into why I went. This place was my time portal to 1931, the year the fire burned and the Forgotten Five died. The men who succumbed to the Waldron Creek blaze perished trying to save trees, timber that might have fed this very mill. This spot was my connection to them, our historical bridge. It was here that I felt closest to their story.

But my link to this place was much more complicated and deeper-seated than its simply being a conduit to the past. This discovery took a little bit longer to arrive at. The timber age was born, it flourished and places like Polley's Mill in Missoula sprang up to process the logs, which Polley's Mill did for nearly a century. By the 1970s, the industry had declined to the point that the mill was no longer viable, and it was closed down. The property, in a sense, passed away, and for the next twenty years or so it remained this way: dead.

But now, after years of quietus, the space is being revitalized as the Old Sawmill District. It's now a baseball field for the local minor-league team, a public park and, in the near future, apartments and condominiums, retail spaces, and other mixed uses. It has a fresh lease on life. The prospect of coming back again as something new. An opportunity to be reborn. Did the Forgotten Five—do any of us for that matter—get a chance to do the same?

At its core, this book has been about unanswered questions, and this is perhaps the grandest of all inquiries: is there an afterlife for us? That answer will come, in due time. Until then, I must simply have faith that it will be resolved in the affirmative. If the answer is, indeed, yes, then I hope the hereafter provides me with a chance to meet the Forgotten Five, to shake

Old Sawmill District. Missoula, Montana, June 2015. *Author's collection.*

their hands, to thank them for their sacrifice and to finally learn what really happened on that hot August day in 1931.

I have yet to see the dark stranger again since that last day when I caught a fleeting glimpse of him in Silver Park, shortly after the guts of the monster were removed. I have a hunch that the next time I meet him, he, too, will hold some of the answers I seek.

NOTES

PROLOGUE

1. "Ladder fuels" is a firefighting term for both live and dead vegetation that allows a fire to climb up from the forest floor or landscape into the tree canopy. It includes tall grasses, tree branches and shrubs.
2. A fireline constructed by hand tools consists of a gap in combustible fuels, cleared down to mineral soil, and is usually about one to one and a half feet in width.

CHAPTER 1

3. Robert H. Frank and Ben S. Bernanke, *Principles of Macroeconomics*, 3rd ed. (Boston: McGraw-Hill/Irwin, 2007), 98.
4. A mucker is a person who shovels broken ore and waste rock into ore buckets or ore cars.
5. D. Roth and G.W. Williams, "The Forest Service in 1905," http://www.docstoc.com/docs/120692778/THE-FOREST-SERVICE-IN-1905---USDA-Forest-Service (accessed September 25, 2014).
6. Clearcutting is a logging practice where most, if not all, of the trees in a given area are uniformly cut down.
7. R.Y. Stuart, *Report of the Forester* (Washington, D.C.: United States Department of Agriculture, Forest Service, 1932), 33.
8. Timothy Egan, *The Big Burn* (New York: Houghton Mifflin Harcourt, 2009), 259.

9. J.B. Davis, "The True Story of the Pulaski Fire Tool," *Fire Management Notes* 47, no. 3 (1986).
10. The actual inventor of the Pulaski tool is subject to debate. A combination axe head/mattock head tool developed by the Collins Tool Company was on display at the nation's Centennial Exhibit in Philadelphia in 1876. That nickel-plated tool later was on display at the Smithsonian Museum of Arts and Industry in Washington, D.C. Pulaski would have been about six years old at this time.
11. Egan, *Big Burn*.
12. M.P. Malone, R.B. Roeder and W.L. Lang, *Montana: A History of Two Centuries* (Seattle: University of Washington Press, 1976).
13. S.V. Cooper et al., *Forest Habitat Types of Northern Idaho: A Second Approximation, USDA Forest Service, Intermountain Forest and Range Experiment Station, General Technical Report, INT-236* (Washington, D.C., 1991); K.F. Kipfmueller and T.W. Swetnam, "Fire-Climate Interactions in the Selway-Bitterroot Wilderness Area," in *Wilderness Science in a Time of Change*, Conference Vol. 5, *Wilderness Ecosystems, Threats and Management, USDA Forest Service Rocky Mountain Research Station, Proceedings RMRS-P-15-Volume 5*, edited by D.N. Cole et al. (Washington, D.C., 2000); M.L. Rorig and S.A. Ferguson, "Characteristics of Lightning and Wildland Fire Ignition in the Pacific Northwest," *Journal of Applied Meteorology* 38 (1999): 1,565–576.
14. Stuart, *Report of the Forester*, 33.
15. A crown fire occurs when the fire advances from top to top of trees, independent of fire on the ground.

Chapter 2

16. The terms "smokechaser" and "fireguard" were often used interchangeably, which could and did lead to some confusion. Both were responsible for putting out fires; however, they were different positions. Typically, smokechasers did exactly what their name suggests, chasing after smoke (fires). Thus, they were primarily firefighters. Fireguards helped to suppress fires, but they had other duties as well, such as serving as lookouts, patrolmen, campground registers, protective assistants or dispatchers.
17. History of Forest Reserves and Forest Rangers, http://www.fs.usda.gov/detail/eldorado/learning/history-culture/?cid=STELPRDB5262329 (accessed November 10, 2014).
18. R.J. Duhse, "The Saga of the Forest Rangers," *Elks Magazine* (July/August 1986): 7.
19. Ken Robison, "The Ozark Club," *Distinctly Montana*, http://www.distinctlymontana.com/montana-places/09/08/2011/ozark-club (accessed November 12, 2014).

Chapter 3

20. *Missoulian*, "Forest Fires Take Another Worker's Life," August 31, 1931.
21. Ibid., "Fires Bring Cheer to Work-Hunters," August 24, 1931.
22. Stuart, *Report of the Forester*, 1.
23. Interestingly, in 1886, soldiers in Troop "M" of the First United States Cavalry under the command of Captain Moses Harris worked as firefighters in Yellowstone National Park. Troops would remain in the park for the next thirty-two years. Little is known about how many fires they fought during this period, if they received any fire training or whether any injuries or fatalities occurred among the soldiers during these firefighting efforts. Although they worked as firefighters when needed, they were soldiers first. Thus, they could not really be considered primary firefighting resources.
24. By 1925, Porterville, California, had organized Forest Service fire crews, but the program was small and limited only to this specific area of the country.
25. Marty Alexander, "The Interregional Fire Suppression Crew," *Fire Management* (Summer 1974).

Chapter 4

26. Forest Service employees refer to the Okanogan-Wenatchee National Forest as the O Bar W for short.
27. P. Hinton-Walker et al., "The Intersection of Patient Safety and Nursing Research," *Annual Review of Nursing Research*, Vol. 24, *Focus on Patient Safety*, edited by J.J. Fitzpatrick and P. Hinton-Walker, 8–9 (n.p.: Springer Publishing, n.d.).
28. K. Schulz, "The Story that Tore Through the Trees," *New York Magazine*, September 9, 2014.
29. A. Neuhauser, "As Wildfires Worsen, Calls for Changes in Tactics," *U.S. News and World Report*, November 5, 2014.

Chapter 5

30. Norman Maclean, *Young Men and Fire* (Chicago: University of Chicago Press, 1992), 152.

Chapter 7

31. The African Methodist Episcopal Church was cofounded in 1816 in Pennsylvania by Richard Allen. Allen was the first black ordained minister in the Methodist Episcopal Church. He opposed the church's

custom of segregating black parishioners to special pews, which they had to relinquish to white members when the remaining seats of the church were filled. In response, he formed the first AME Church, which had to meet in his Philadelphia blacksmith shop until a more proper house of worship could be constructed.

Despite growing up in Great Falls, I had never really noticed or been aware of the Union Bethel AME Church. Union Bethel, I came to learn, is one of the earliest-built, longest-used churches for African Americans in Montana. The original church, a one-story wooden affair, was first built in 1891 on land sold by Great Falls founder Paris Gibson. By 1916, the church and its wooden components were deteriorating to such a point that, with the assistance of the Greater Great Falls community, renovations were undertaken. The "new" church, completed in the Gothic Revival architectural style, was finished by 1917. In a letter to the editor published in the *Great Falls Tribune Daily*, the Reverend W.H. Prince wrote, "The remodeling of our church…means much for the moral and social life and advancement of our people." On September 11, 2003, Union Bethel was placed on the National Register of Historic Places. K. Ogden, "Historic Black Church Shines with New Light," *Great Falls Tribune*, n.d., http://archive.greatfallstribune.com (accessed April 18, 2015).

Chapter 10

32. The 82nd Division was formally activated in August 1917. A contest was held to give a nickname to the new division, and "All American" was selected since the division was composed of soldiers from all forty-eight states. In August 1942, the 82nd became the army's first airborne division, and its name was changed to the 82nd Airborne Division.

INDEX

C

D

E

F

G

M

N

O

P

ABOUT THE AUTHOR

Charlie Palmer worked for nearly twenty years as a wildland firefighter. He spent ten years of his fire career with the Missoula Smokejumpers, based in Missoula, Montana.

Between fire seasons, Charlie completed his BA, MA, EdS and EdD degrees at the University of Montana in Missoula. In August 2006, Dr. Palmer joined the University of Montana's Department of Health and Human Performance, where he continues to teach and conduct research, primarily focusing on human factors in wildland firefighting and other high-risk occupations.

Since 2000, Charlie has served as a consultant with Mission-Centered Solutions, helping to teach the Point of the Spear (L-380) and the All Hazard Incident Leadership (L-381) courses around the world. He also currently sits on the board of directors and is the director of human performance for the First Twenty, a national nonprofit organization dedicated to improving the health of American firefighters. Charlie also serves as historian for the National Smokejumper Association.